101
Best Aviation Attractions

Other Books by John Purner

The $100 Hamburger, Second Edition
The $500 Round of Golf

101 Best Aviation Attractions

John Purner

McGraw-Hill

New York Chicago San Francisco Lisbon London Madrid
Mexico City Milan New Delhi San Juan Seoul
Singapore Sydney Toronto

The McGraw-Hill Companies

Library of Congress Cataloging-in-Publication Data

Purner, John.
101 best aviation attractions / John Purner.
p. cm.
Includes index.
ISBN 0-07-142519-5
1. Aeronautical museums—Directories. I. Title: One hundred one best aviation attractions. II. Title.

TL506.AIP87 2004
629.13'0074—dc22 2004040262

1 2 3 4 5 6 7 8 9 0 DOC/DOC 0 9 8 7 6 5 4

ISBN 0-07-142519-5

The sponsoring editor for this book was Scott Grillo, the editing supervisor was Caroline Levine, and the production supervisor was Pamela Pelton. It was set in Perpetua by Wayne A. Palmer of McGraw-Hill Professional's Hightstown, N.J. composition unit. The art director for the cover was Margaret Webster-Shapiro.

Printed and bound by RR Donnelley.

This book is printed on recycled, acid-free paper containing a minimum of 50% recycled, de-inked fiber.

Contents

Acknowledgments

I love hot dogs but know better than to watch them being made.

Writing a book is a lot like making a hotdog. It isn't pretty work. Sometimes what comes out the other end is as enjoyable as a hot dog, but it sure is messy work. Scraps of paper with precious notes wind up everywhere. Hard disks crash and tempers flare. Each day the mailman brings more material that might make its way into the book from some faraway press agent for a museum. Some telephone calls are returned while others fall on deaf ears.

To grind words and ideas into a book you must have an understanding yet gently pushy publisher. I have the best one in the world—McGraw-Hill's Scott Grillo. He came to his lofty position in that Manhattan high-rise honestly—he earned it! An email comes from Scott infrequently but always at just the right time. It is his reminding and remembering that gets books written. Especially this one! Thanks Scott.

I owe this book and everything else to Renée. Without her it would never have been written. She was always there with encouragement and direction. As busy as she is, Renée found time to become my researcher and keep the madness that is book writing sanely organized. There was always a light at the end of the tunnel. Thankfully, it was Renée with a flashlight guiding me forward and not the freight train that I truly deserved. Thanks sweetheart for helping me to become what I longed to be. Forever and two Wednesdays!

Introduction

101 Best Aviation Attractions is a book about flying written for people who are as passionate about airplanes as I am. Airplanes are to be flown, studied, and admired as are the people who design, build, and fly them.

This book is about museums, air shows, hallowed ground, and adventure. Please notice that I did not rank the *101 Best*. They are listed alphabetically in the Contents. Each is wonderful and very worth visiting. Certainly, honest people can debate why I left out a certain place of importance and included something that many may feel is not very important at all. I selected the venues that have been the most important to me and hope that you will enjoy them. Debate is healthy, though, so I have created a website as a companion to this book. The URL is www.flyingsbest.com. I hope you will come to it often and make your comments about the places I selected, blast me if you must, and complement me if you can. Next, please suggest other places that you feel are just as important as the ones in this book or that you feel are even more important. Finally, and most importantly, please send me the story of your trip to the various spots in this book and your experiences with the adventures I suggested. I would really like to hear from you, and I know that the other aviation afficionados who visit this website will enjoy sharing with you.

While I do feel that each of the *101* is a giant and should not be missed, I can't help but have a favorite or two and an idea of which stands a little taller than the rest. Here goes! When it comes to museums my pick is the **National Museum of Naval Aviation** in Pensacola, Florida. The reason is simple. When you go there you will find an excellent museum that is very well run with a *Big Plus* feature that tips the scale. The Blue Angels base there and practice there and, if you time your visit with only slight care, you will see them practice after you tour the museum. A wonderful museum and a flying exhibition by the best flight demonstration in the world are very hard to top. When you consider the price of admission, *free*, it is unbeatable!

If I had to pick only one hallowed site to visit, it would be a department store parking lot on Long Island. Here, there is a large plaque that lets us know that in 1927 this wasn't a department store parking lot—it was part of Roosevelt Airfield. Charles Lindbergh began one of the two most important flights in the history of aviation from here. His wheels left the ground at exactly the spot where the plaque has been erected. To stand here is to be reminded that one man can make a difference. In a very real way it is always only one man that does make a difference. On May 20, 1927 the man was Charles Lindbergh and the difference he made was monumental.

In this book you read about many aviation adventures that you can take. Each is amazing! If you can do them all, go for it. One is more important to me than the others; it is the chance to fly on a B-24. My father was a member of a B-24 crew in World War II and so was my uncle, his younger brother. I have spent good hours talking with my Uncle John about his life inside the belly of a B-24 and his memories of being shot out of the sky one fateful day over Berlin. I have read my father's war diary and have been thrilled by it. I have stood next to a B-24 or two and crawled around inside of one with my then 2-year-old daughter. What I have never done, but will someday soon, is go flying in one. I want to tell my Uncle what I felt. That's the adventure I have picked out for me. Which will you choose and when you will go? Please send me an email and a picture or two; I'd love to know what you did and how it went. My email address is jpurner@flyingsbest.com.

I am an air show lover. I like to hear the engines and see the pilots make their planes do what my Cherokee can't—maybe it could but not while I'm flying it. Someday I want to go back to Texas to **Harvey and Rihn**, for if there is a place to learn to control an airplane, it is there.

A few years ago, I once in a while flew an Ercoupe I owned to Lake Jackson, Texas, and watched Debby Harvey-Rihn and her husband "Doc" Harvey turn an Extra 200 every which way but loose. Today Debby continues to run the business they built together and to compete. She is a world champion at both. I know that her people can make even me a better pilot. We'll see!

What air show is the best for me? Without question it is the **Mid Atlantic Air Museum's World War II Weekend Airshow** in Reading, Pennsylvania. This is living history at its best—an airshow with 1,000 actors dressed in period costume in a town, decorated for the occasion, with an event to delight everyone. If you want to learn about World War II flying and flyers and the girls they left behind, head for Reading in early June.

If I had just one special place to fly to with Renée this year, where would I go? For me that's easy. I'd go back to Lake Geneva to the **Grand Geneva Resort and Spa**. We have been there before and, hopefully, will go again this year when we're back in Chicago. There are some fly-in resort attractions in this book that I hope will find favor with you; they have with me.

What's the best way to use this book? Read it all—one time. Then have another look with a pencil in one hand and a stack of post-its in the other. Write some notes in the margins and mark some pages. Make a plan and follow it through. Wherever you decide to go, be sure to take someone along and share the moment. Memory-making is so much better that way.

#1

Abbotsford International Airshow

Abbotsford is just across the U.S.–Canadian border near Bellingham, Washington. It is an easy drive from Bellingham if you chose to come up from the Seattle area. You can certainly count on the Snowbirds performing every year as well as other Canadian acts not seen in the United States. Canada is bush pilot country. Keep your friendly smile on and meet as many people as you can. Sooner or later you'll be in the presence of a Canadian who can talk your arm off about flying in the back country. These people use airplanes as a tool. Planes are part of their everyday life up there.

The thing I like best about this show, and I think you will too, is the camping opportunity. If you fly in, you may camp under the wing of your aircraft right on the show property. If you have driven up from the United States with an RV there is a special campground that is set up just for the airshow. It is on the airport, and a shuttle operates continually during the show to bring you to the viewing area. Plan to arrive Thursday and depart Monday. A 20- by 50-foot campsite costs only $135.00 for the entire show, and show tickets are included in that price! Do it this way and you'll maximize your show experience by spending 4 days living with other fans.

When the show ends and it's time to leave, *don't!* You're a short drive from Vancouver City and only a ferry ride from Vancouver Island. Both are knock-your-socks-off gorgeous. See them!

Dates

August 13, 14, 15, 2004

Hours

Sunrise to sunset

Admission Cost

Adult	$20.00
Child (6–12 years)	$10.00
Child 5 and under	Free
Carload (not to exceed 8)	$75.00
Runway seating (in addition to any other admission)	$10.00

Prices include parking and Goods and Services Tax

Contact Information

Abbotsford International Airshow Society
1276 Tower Street, Unit #4
Abbotsford, British Columbia
Canada V2T 6H5
(604) 852-8511
(604) 852-6093 (fax)
info@abbotsfordairshow.com
http://www.abbotsfordairshow.com

G/A Airport Serving the Event

Abbotsford, B.C., Canada (Abbotsford Airport—CYXX)
(604) 855-1001
(604) 855-1066 (fax)
http://www.flyabby.com

While slot times are not required, they are recommended. Phone (604) 556-3303 to get a transponder code and a slot time. Please specify whether you require day or overnight parking when making your reservation. You must arrive no later than 10:15 a.m. Aircraft arriving for the day will be parked in a special day parking area. Fly-in guests are welcome to pitch a tent by their aircraft and spend the night or the weekend! Take advantage of the free shuttle service to meet fellow flyers at the **Abbotsford Flying Club!** Overnight aircraft parking will be close to the campground area allowing easy access to amenities. To ensure overnight aircraft security, access to overnight aircraft parking will be restricted to pilots, their passengers, and approved guests. You can prepurchase tickets and camping space through the airshow office by telephone at (604) 852-8511. Visa and MasterCard are accepted.

FBO Name/Phone Number

Copac Air Services (Shell)
(604) 854-1964

Godspeed Aviation (24 hours, self)
(604) 854-0887

Ground Transportation

Enterprise Rent A Car
(800) 736-8222

Best Place to Stay $$$

Best Western Rainbow Country Inn
43971 Industrial Way
Chilliwack
British Columbia, Canada, V2R 3A4
(604) 795-3828

#2

Aerospace America International Airshow

Aerospace America International Airshow has twice been named the "Best Airshow in North America" by separate organizations. It has twice been named the "Number One Public Event" in Oklahoma. It has won many awards through the years. In the airshow industry it is known for its many firsts. This is my favorite air show and it will be yours as well. The differences are subtle but they come through loud and clear. It starts with a "we're glad you're here" attitude, which is not so evident everywhere.

Let me give you an example. Oklahoma is an aviation state and has always been an aviation state. Oklahomans are compelled to sell everyone on aviation, particularly general aviation. To that end they want as many pilots to fly in for the show as possible. That is not the case at most airshows. Attendees are often discouraged from landing in their own aircraft. Here they actually give you *free* admittance if you show up in an airplane. These folks love airplanes! That's why this is such a good show.

The **Aerospace America International Airshow** will be headlined by the Blue Angels as will 36 other air shows this year alone. They are my personal favorites and I try to show up wherever they are flying. The Blues will only be at the show on Saturday and Sunday. There will be about a dozen other top aviation performers here. All the acts are jaw droppers. Additionally, a few armed forces aircraft will show up for quick flybys—maybe a B-2 perhaps an F-117.

On the ground you will be treated to almost 100 static displays of some of the greatest warbirds you have ever imagined.

If you can come to only one air show this season. This is it!

Dates

June 18–20, 2004

Hours

Friday	4:00 p.m.–9:30 p.m.
Saturday	9:00 a.m.–5:00 p.m.
Sunday	9:00 a.m.–5:00 p.m.

Admission Cost

Adults	$15.00
Children (6–12)	$6.00
Children (under 5)	Free

Contact Information

Aerospace America Team
(405) 685-9546
cgdon@aerospaceokc.com
http://www.aerospaceokc.com

G/A Airport Serving the Event

Oklahoma City, OK (Will Rogers World—OKC)
Phone: (405) 680-3200
Tower: 119.35
AWOS: (405) 682-4871
Runways:

13–31	7800′ × 150′
17L–35R	9802′ × 150′
17R–35L	9800′ × 150′
18–36	3079′ × 75′

What better way to attend an airshow than to fly in in your own airplane? Beat the traffic, and show the public that they, too, can learn to fly and own their own airplane. Show management will even bribe you. Bring your airplane and you won't have to purchase a ticket. You get to see Aerospace America for free!

Check NOTAMS prior to departure for the airshow. McAlester Flight Service Station's new Mobile FSS will be on the field. Be sure to visit them to get a briefing or file your flight plan prior to departure. The airshow flight operations staff will also brief all pilots on proper departure procedures.

Plan to arrive by 3:00 p.m. on Friday and by 8:00 a.m. on Saturday and Sunday. When the show starts the airport will be closed. No arrivals or departures will be permitted until after the final performance each day.

Parking will be on grass, except for large, heavy aircraft, so bring sturdy tiedowns to secure your plane in the Oklahoma wind.

Mike Grimes
Fly In Chairman
(405) 425-2017
mgrimes@dps.state.ok.us

FBO Name/Phone Number

AAR Oklahoma
(405) 218-3000

Ground Transportation

Rental car

Best Place to Stay $$$$

Renaissance Oklahoma City Convention Center Hotel
10 North Broadway
Oklahoma City, OK 73102
(405) 228-8000

Best Place to Stay $$$

Hilton Garden Inn Oklahoma City Airport
801 S. Meridian
Oklahoma City, OK 73108
(405) 942-1400

#3

Air & Sea Show

The annual **McDonald's Air & Sea Show** attracts more than four million people and is one of the world's largest spectator events. It takes place along 4 miles of Fort Lauderdale's oceanfront. Top military and civilian pilots perform each year.

Annually, the men and women in the military are the guests of honor at many parties and special events that are open to the public during **Fleet Week USA.** The week concludes with the huge **McDonald's Air & Sea Show,** a two-day event. The public pays no admission fee to see the show, thanks to corporate support from 40 sponsors.

Fort Lauderdale boasts one of the best beaches anywhere. For decades it has been the epicenter of the collegiate spring break, and with good reason. The weather is year-round perfect. A little warm in the summer perhaps but to die for in the winter.

Let your imagination run wild for a moment: sun, beach, gorgeous ocean, the best aircraft, with the best pilots the world has ever known flying overhead at the speed of heat. Add just a pinch of oceans-borne performers, throw in a dab of amazing sunsets, stir in a get-your-heart-pumping fireworks show and what you have is a recipe for a very good time. That's exactly what **McDonald's Air & Sea Show** is, a very good time.

Condos and hotels line the beach from Miami to well north of Fort Lauderdale. Finding a satisfying place to stay is not going to be the problem. Affording it is! Fort Lauderdale is everybody's idea of a great escape, and there is a limited amount of beachfront property so the cost of ownership is high and passed straight down the food chain to vacationers like us.

Miami and Fort Lauderdale have earned their title as "the capital of the Caribbean" through the kitchen. The best island-borne cuisine is available. You'll want to make reservations. Remember people come here from all over the world to relax and enjoy. **Fleet Week USA** is especially crowded. Restaurant prices can be high.

If you're in the mood for a little nightclubbing, head to Miami's world-renown South Beach. You'll have an evening that won't soon be forgotten. These folks know how to have a good time.

Vacationing in Miami and Fort Lauderdale doesn't have to be expensive. Doing it all on a very modest budget takes some work and planning, but it can easily be done.

Airline fares into Miami International are good deals. The Fort Lauderdale/Hollywood airport is closer to where you are going but there are fewer flights and higher fares; one goes with the other. *Book early!*

Rent a car. Rental cars in California and Florida are a thoughtful vacationer's bonanza. A compact car will run you just a bit over $100 a week with unlimited mileage. *Book early!*

You'll need a rental car because a man on a budget doesn't stay on the beach at $250 a night. Move inland and find great deals on golf course–owning resort hotels for less than $100 a night. With a little looking you can find a great room at a national motel chain for less than $50 a night. *Book early!*

What about low-cost meals. Florida is the right state for low-cost vacationing. Many of the older hotels and motels have kitchenettes in the rooms and barbeque pits on the patios. One stock-up trip to the local grocery store will cost less than a single meal in a restaurant. A lot less! Look for a room with a kitchenette and you'll find plenty of money in your pocket when you get home.

One hint about the rental car. Just because you have it doesn't mean you have to drive it. Parking will be scarce near the viewing area. Visit the shows Website and get the most current schedule for the shuttle buses. Find a hotel within walking distance of one of the pickup points. The cost will be *free* to maybe $2. Save the car for a trip to South Beach. Go and see it at least!

Dates

Annually during the first week of May

Hours

The airshow will happen all day long.

Admission Cost

Free

Contact Information

Show Hotline: (954) 527-5600 Ext. 4

www.airseashow.com

www.fleetweekusa.com

E-mail

Check the Website. There are different e-mail addresses for each area of concern.

G/A Airport Serving the Event

Fort Lauderdale, FL (Ft. Lauderdale Executive—FXE)

Phone: (954) 938-4966

Tower: 120.9

ASOS: (954) 772-2537

Runways:

13–31 4000′ × 100′

8–26 6000′ × 100′

Full service is offered at FXE and private aircraft will be much happier here than at Ft. Lauderdale/Hollywood—FLL. While both are great airports with ILS approaches, FLL has many, many, many airliners taking off and landing. If you decide to go in there, the controller will repeat my warning, *"Caution wake turbulence!"*

FBO Name/Phone Number

Banyan Air Service, Inc.

(954) 491-3170

www.banyanair.com

info@banyanair.com

Southeast corner of the field.

If you're flying a 100LL burning single or twin, Banyan Air Service, Inc. is definitely where you want to be.

Ground Transportation

Enterprise Rent A Car

800 Rent A Car

Best Place to Stay $$$$$

Harbor Beach Marriott Resort & Spa
3030 Holiday Drive
Fort Lauderdale, FL 33316
(954) 525-4000

The Harbor Beach Marriott Resort & Spa is tucked away on 16 sun-drenched oceanfront acres. It boasts an irresistible list of indulgent pastimes, superior meeting space, and outstanding service. It is also close to the chic dining places, boutiques, and art galleries of Fort Lauderdale's celebrated Beach Promenade and Las Olas Boulevard…and therefore close to perfection. Bring the kids, activities have been planned with them in mind.

Best Place to Stay $

La Quinta Inn and Suites–Ft. Lauderdale Plantation
8101 Peters Road
Plantation, FL 33324
(954) 476-6047

The La Quinta Inn is located just minutes from popular attractions, including Fort Lauderdale Beach. It features popular amenities, including fitness centers, heated pools, and spas. It has earned a AAA three-diamond rating.

Best Aviation Attraction

#4

Air Combat USA

Walter Mitty lives in the deep recesses of all of our minds. The 6 o'clock news awakens my favorite pretend. I close my eyes and become a fighter pilot. Maybe you do, too. If you're young enough and motivated enough, you can turn that dream into a life. For most of us it will always be merely a dream. Television and the movies feed into it. For brief moments we identify a character and become that person. But seeing and imagining is not the same as feeling, touching, smelling, and experiencing all that could have been but never will.

If this is your dream and you have the courage and dollars to go along with it, you can for a day, become a fighter pilot. You can for a day, shoot at an adversary as you try to avoid the blasts of that person's laser gun. How is this possible? Where do I go? Who do I call? Those were my only questions and I'll bet they're yours as well.

All across the country small companies have sprung up to satisfy our need to become a fighter pilot for a day. There is probably one within a 2-hour drive of your home. That is the case for me. But for me, I want to fly only with the best. I want to go up with people who flew real fighters in real wars and taught real military men how to become as good as they are. I want to fly in equipment that will make me feel that I am indeed in a jet fighter. I want to fly with a group that does this frequently and successfully. I am not into being a guinea pig. Nor do I want to feed into someone else's fantasy of making a million dollars in an aviation business. I want to fly with someone who has already made it.

Air Combat USA in Fullerton, California, is about as far away from my home in Orlando as you can get and still be in the United States! It is a place where you or I can actually fly a light attack fighter aircraft and become a fighter pilot for a day.

First you'll be coached by a "real" air combat instructor. Someone who has actually taught military pilots the fine art of real air combat. Next you'll climb into the cockpit of a SIAI Marchetti SF260 with your instructor by your side and maneuver through a series of dogfights. This is not a simulator. You're controlling and riding 8000 feet in the sky in a military trainer. No pilot's license is required for this flying adventure.

Over 30,000 people have been a Fighter Pilot for a Day since Air Combat USA began operating in 1989. Here you'll learn and practice the tactics and maneuvers taught by the military, then you'll fight to win in real aerial combat. For many this will become the most memorable experience of their life.

There are other air combat reality vendors, but there are none better and I can't think of a single one that is even half as good. Air Combat USA has more pilots on its payroll than its competitors have customers at the cash register. They are the heavy hitters of this game. This is the big league!

Safety is the first thing to consider when you sign up for one of these adventures. Will you survive the experience? Safety is not just a phrase or an attitude; it is the result of countless activities and many choices. Choice number one is the aircraft. You want one that is truly aerobatic and will simulate air combat. Most of all, you want one that won't come apart if flown anywhere close to properly. The SIAI Marchetti SF260 fills the bill. It is a complex high-performance aircraft that is both stable and predictable. It is suitable for +6 and −3 G's and although it is a military trainer, it behaves like most general aviation complex aircraft. Unlike the T-34 flown by most competitors, I can't think of a single in-flight structural failure attributed to this bird. I am not saying that it hasn't happened, only that I have never heard about it. Sadly, this is not the case with the T-34. Just last year (2003), one came apart near Houston, Texas, while being flown by a very experienced pilot. Both he and his customer did not survive the crash. You shouldn't have to pay for your flying adventure with your life. Yes, you'll wear a parachute just in case!

Adventures are best shared. No, you shouldn't take your wife and kids along for the ride. They won't fit in the cockpit anyway. Each aircraft is equipped with three onboard video cameras. This is one adventure that's never really over. You can relive it for years to come and share it with your family and friends. They won't just hear what you did; they'll actually see you do it!

Yes this is an expensive day. The cost varies based on the program you choose.

Phase I: Basic Air Combat Maneuvers	$ 995/Person
Phase II–IV: Intermediate through Advanced (ACM)	$ 995/Person/Phase
Fighter Lead In Program	$1695/Person
Advanced Fighter Tactics	$1695/Person
Air Combat USA Fighter Squadron (Members Only)	$ 500/Person

Each of these programs is fully explained in the Air Combat USA brochure and Website. Some require that you have participated in a previous course, others don't. Yes, it is expensive but what is the most memorable adventure of your life worth?

Fullerton Municipal Airport, the home of Air Combat USA, is only 6 miles from Disneyland. Combine this adventure with a family trip to the House of Mouse and everyone will be happy!

If you can't find the time for a getaway to Fullerton don't put this dream out of your mind. In 2004, Air Combat USA will take its show on the road. They'll be in Chicago, Dallas, New York, Seattle, and a few more cities for a total of eleven.

Keep the dream alive. Do this!

Dates

You pick!

Hours

Daylight. Classroom training begins at 0800!

Admission Cost

See above.

Contact Information

Air Combat USA, Inc.
230 N. Dale Place
Fullerton, CA 92833
(800) 522 7590
(714) 522 7590
(714) 522 7592 (fax)
aircombat@pacbell.net
http://www.aircombat.com

G/A Airport Serving the Event

Fullerton, CA (Fullerton Municipal Airport—FUL)
Phone: (714) 738-6323
Tower: 119.1
ASOS: (714) 870-1372

Runway: 6–24 3121′ × 75′

FBO Name/Phone Number

Call for permission to taxi to and park on the Air Combat USA Ramp.

Ground Transportation

Rental car

Best Place to Stay $$$$$

The Ritz-Carlton, Laguna Niguel

One Ritz-Carlton Drive
Dana Point, CA 92629
(949) 240-2000
(949) 240-0829 (fax)

#5

Air Force Flight Test Center Museum

Edwards Air Force Base in California is the birthplace of supersonic flight and the jet age. The first jet aircrafts were developed at other sites but it is here that they were perfected. The history of Edwards 6-decade-long run as the world's premier flight testing and flight research center is preserved and displayed in the **Air Force Flight Test Center (AFFTC) Museum.** Why did it happen here?

Designers and pilots about to make a "first flight" are drawn to the high desert because of its remoteness, the remarkably good flying weather, and the uncrowded skies. The really special thing about this location is the safety afforded by the vast expanse of Rogers Dry Lake. The lake bed has, again and again, served as an emergency landing field. It is the perfect place to take off because it is the perfect place to fly and to land when things go wrong. All of these resources, when combined, make Edwards the optimum location for the first flights of high-performance and experimental aircraft.

In October 1947, the Bell X-1 was piloted through the speed of sound. Soon it was followed by other planes being flown through other unexplored regions of the laboratory in the sky above this high desert. For the first time, a craft would pass Mach 2, 3, 4, 5, and 6, and climb above 100,000, 200,000, and 300,000 feet into near space.

To afford civilians an opportunity to learn about the test center, a museum was opened in 1994. In July 2000, it was moved to a larger more usable facility. Exhibits are crafted to tell the story of the first military uses of Edwards, flight testing during World War II, breaking the sound barrier, aviation records set, testing the X-15, and the life of Glen Edwards, for whom the base was named. Other exhibits include aircraft hardware, personal memorabilia, aviation fine art and photography, and models.

A few "special" aircraft are displayed inside the museum building. The F-16B, an NA-37B, an AQM-34 Firebee drone, the X-25B concept demonstrator, and a full-scale replica of the Bell X-1 are all in perfect condition. Museum visitors can learn about the past, present, and future of flight testing in the museum's theater, and souvenirs can be purchased in the gift shop.

Currently, 16 of the museum's aircraft are displayed outdoors in the museum's airpark. They include the B-52D, CH-3E, F-84F, F-101B, F-104A, F-105D, F-111A, H-34, NF-4C, SR-71A, T-28B, T-33, T-39, UC 45, YA-7D, and Gloster Meteor. The F-16B, F-86E, F-104A, NF-104A, P-59B, T-33A, and T-38A are mounted on pedestals around the base. Eventually, the first T-46; prototypes of the A-7F, A 9, A-10B, F-4E, and F-94; one of only two PA-48 Enforcers; the first production C-141; and a number of other one-of-a-kind aircraft will be added to the exhibit space.

Dates

Currently: Due to security restrictions, the museum is only open to those individuals with official business at Edwards AFB. This condition will soon change. Stay updated by visiting the Website frequently.

Normally: Tuesday through Saturday (closed on federal holidays)

Hours

9 a.m. to 5 p.m.

Admission Cost

Free

Website

http://www.edwards.af.mil/oh.html

Contact Information

Air Force Flight Test Center
Public Affairs
AFFTC/PA
1 S. Rosamond Blvd.
Edwards AFB, CA 93524
Public Affairs: (661) 277-3510

G/A Airport Serving the Event

Mojave, CA (Mojave Airport—MHV)
The Nation's Civilian Flight Test Center
Website: http://www.mojaveairport.com/
E-mail: stuart@mojaveairport.com
Phone: (661) 824-2433
CTAF-Unicom: 127.6
Runways:

12–30	9500′ × 200′
8–26	8000′ × 100′
2–22	4700′ × 50′

Arresting gear is installed on both ends of runway 12-30! Transient aircraft tiedowns are available and highly recommended due to the robust winds adjacent to the airport administration building under the old tower. There is no tiedown fee.

FBO Name/Phone Number

East Kern Airport District
(661) 824-2433

Ground Transportation

Enterprise Rent A Car
800 Rent A Car

Best Place to Stay $$

The Mariah Country Inn and Suites
1385 Highway 58
Mojave, CA 93501
(866) 627-4241
http://www.mariahhotel.com/
reservations@mariahhotel.com

This is a AAA three-diamond-rated motel. It is the nicest facility for many, many miles. Naturally, they have a pool, spa, and workout room. They operate a shuttle to and from the airport. The Mariah Inn restaurant, which is attached, is one of the best in town.

Best Place to Stay $

Econo Lodge
2145 SR 58
Mojave, CA
(661) 824-2463
http://www.econolodge.com/

This is a AAA two-diamond-rated facility. It is clean, comfortable, and affordable. There is a pool and pets are accepted. The bonus feature is that each room is equipped with a microwave and refrigerator.

#6

AirVenture Museum

Oshkosh, Wisconsin's **EAA AirVenture Museum** has become one of the world's most extensive aviation attractions, and a year-round family destination. It is located on the site of the world's largest aviation event, **EAA AirVenture.**

The collection of historic artifacts was begun in 1962 when Steve Wittman donated his famous air racer *Bonzo.* It now comprises more than 20,000 aviation objects of historic importance. Included are 250 historic airplanes, and the count grows almost weekly as exhibits are constantly added. Everything from a powered parachute to a B-17 Flying Fortress is maintained in airworthy condition! Some are available to give ordinary people the chance to fly in historic aircraft. Most are used to provide flight demonstrations or support special activities such as the EAA Young Eagles program.

The **EAA AirVenture Museum's** library contains almost 9000 volumes. The collection covers a variety of topics including biographies, aerodynamics, history, fiction, aeronautics, air racing, and home-building.

The library's photographic collection archives more than 100,000 images of aircraft, spacecraft, and the people who made and flew them. Many photos chronicle the home-building movement of the early 1950s. They tell the story of an emerging group of people determined to design, build, and fly their own aircraft. Important photo archives donated by private collectors are curated here. Included are:

The Radtke Collection
A thousand negatives of military aircraft, civilian aircraft, and famous aviators from the 1930s.

The Worthington Collection
More than 125 glass negatives taken by an unknown photographer.

The Zeigler Collection
More than over 200 glass negatives of early German aviators of the post–World War I era.

The Norman Collection
Hundreds of 8- by 10-inch black and white photographs covering the golden years of aviation.

This is truly one of the great aviation museums. Do not pass up an opportunity to visit. Be warned! Timing your visit to coincide with the EEA's annual airshow is a great mistake. It becomes very crowded and is hardly "user friendly" at those times. Plan your trip for a nice spring or summer weekend. Combine your trip with a visit to the **Pioneer Airport,** which is right next door. Here you want to spend thoughtful time and soak up the history that is all around you. Plan to spend time here.

Dates

Open year round

Closed New Year's Day, Easter Sunday, Thanksgiving Day, and Christmas Day

Hours

Monday–Saturday	8:30 a.m.–5:00 p.m.
Sunday	10:00 a.m.–5:00 p.m.

Admission Cost

Adults	$8.50
Seniors	$7.50
Students (8–17)	$6.50
Children (7 and under)	Free
Family Rate	$21.00

Contact Information

EAA AirVenture Museum
PO Box 3065
Oshkosh, WI 54903-3065
(920) 426-4818
museum@eaa.org
http://www.airventuremuseum.org

G/A Airports Serving the Event

Oshkosh, WI (Wittman Regional Airport—OSH)

Wittman is a truly wonderful general aviation airport. It offers four runways, the longest of which is 8002 feet. Every IFR approach including an ILS is available.

Looking for something to do with your airplane? The Museum makes Oshkosh an interesting and convenient fly-in destination, all year round. Free shuttle service is available during Museum operating hours by parking at **Orion Flight Services, Inc.**

FBO Name/Phone Number

Orion Flight Services, Inc. (OSH)
(866) 359-6746

Ground Transportation

City	Rental Car Company	Phone
Oshkosh	Hertz Rent-A-Car	(800)654-3131
	Avis Rent-A-Car	(920)730-7575
Appleton	Hertz Rent-A-Car	(800)654-3131
	National Car Rental	(920)739-6421
	Enterprise Rent-A-Car	(800)RENT-A-CAR

Best Place to Stay in Oshkosh $$$

Hilton Garden Inn Oshkosh
1355 West 20th Avenue
Oshkosh, WI 54902
(920) 966-1300
(920) 966-1305 (fax)

This premier hotel is right on the Wittman Regional Airport grounds next to the **EAA AirVenture Museum.** Plan on spending 2 days here, you'll need at least that much time!

Best Place to Stay in Appleton $$$

Hilton Garden Inn Appleton
720 Eisenhower Drive
Kimberly, WI 54136
(920) 730-1900
(920) 734-7565 (fax)

Best Aviation Attraction

#7

Albuquerque International Balloon Fiesta

After centuries of staring at the clouds man's dream of flight was first realized in France in 1783. On November 21 of that year, Jean-François Pilâtre de Rozier and François Laurent, Marquis d'Arlandes, became the first human pilots of an untethered balloon. Their flight was made possible by the experiments of Joseph and Etienne Montgolfier who began working with hot air balloons earlier in the year. On September 19, the Montgolfier brothers had conducted a demonstration of flight at Versailles. A sheep, a duck, and a rooster became the first hot air balloon passengers.

What they began is today celebrated in a brightly colored extravaganza held each year in New Mexico. More than 750 balloonists participate. The featured and highly photographed event is the mass ascension. All 750 balloons launch from Balloon Fiesta Park just outside Albuquerque. It takes less than 2 hours to get them airborne and it is truly a magical moment. Can you imagine the sound of 750 high-volume gas burners touching off all at once? The roar is reminiscent of a volcanic eruption. Can you imagine the sky filled with 750 colorfully designed nylon envelopes? The scene is not so much a sight as it is a spectacle. The sky is filled with balloons. Bring your camera and don't forget batteries and film.

There are many events during this 9-day show. Each day starts at 5:30 a.m. with the **Dawn Patrol.** The air is smooth in the morning and the pilot participants can't wait to take advantage of it. Your day will end at 8:00 p.m. with the hour-long **AfterGlow Fireworks Show.** Along with the fireworks you'll see several balloons fire up their burners. The glow of the fire through the nylon envelope at night is amazing. Each looks very much like a huge Christmas tree ornament.

Most of the events happen in full view of the Fiesta Park spectators. Most involve hot air balloons. **America's Challenge** is an exception to both rules. It is a long-distance *gas balloon* race. Like every other event, the balloons lift off from Balloon Fiesta Park. The winning balloons will fly for 3 or more days. Eventually they'll land somewhere on the east coast of the United States.

The Balloon Fiesta Park covers 360 acres. The launch field alone is 80 acres. Here's the best part—because of an unblemished 32-year safety record, this is the only balloon event where spectators are actually able to walk among the balloons and talk to the pilots before and during a launch.

It is a hands-on event, organized entirely by the Albuquerque International Balloon Fiesta, Inc. It is remarkable that this nonprofit organization is able to pull it off year after year with only 10 full-time staff members. Tickets are available through their Website www.balloonfiesta.com, or by calling 1 (888) 422-7277 or going to the Balloon Fiesta's Gift Shop at 4401 Alameda Blvd. NE if you happen to be in Albuquerque. The best deal is the All Event Ticket for $50. Buy it and you're set for the event, except for parking.

There is a Preferred Parking pass, which costs $45 for the entire week. General parking lots on the north and south ends cost $5 every time you park. Each parking lot is color-coded and the gates are numbered. This makes finding your car a lot easier.

If you would like to arrange a balloon ride, contact Rainbow Ryder, Official Balloon Ride Operator at (505) 293-0000, 1 (800) 725-2477, fax (505) 237-1339, e-mail: morehotair@rainbowryders.com.

Dates

This is an annual event normally held in October. It will next be held October 2–10, 2004.

Hours

Each day begins with the 5:45 a.m. launch of the **Dawn Patrol Show.**

The days end at 8:00 p.m. with the hour-long **AfterGlow Fireworks Show.**

Admission Cost

The All Event Pass is $50.00.

Tickets for individual venues can be purchased for $5.00.

Website

http://www.balloonfiesta.com

Contact Information

Albuquerque International Balloon Fiesta
4401 Alameda Place NE
Albuquerque, NM 87113
Toll-Free: 1 (888) 422-7277
Phone: (505) 821-1000
Fax: (505) 828-2887

E-mail

balloons@balloonfiesta.com

G/A Airport Serving the Event

Albuquerque, NM (Albuquerque International Sunport—ABQ)

http://www.cabq.gov/airport/

FBO Name/Phone Number

Cutter Aviation ABQ, Inc.

www.cutteraviation.com

(505) 842-4184

(800) 678-5382

Ground Transportation

Enterprise Rent A Car

(800) 736-8222

Best Place to Stay $$$$$

Hyatt Regency Tamaya Resort & Spa

1300 Tuyuna Trail

Santa Ana Pueblo, NM 87004

(505) 867-1234

(800) 633-7313

This hotel is ranked as one of the top 75 resorts in North America by the readers of *Condé Nast Traveler.* You'll love this place and you will be shocked by the great deals.

Best Place to Stay $

Super 8 Motel—Rio Rancho

4100 Barbara Loop Se

Rio Rancho, NM 87124

(505) 896-8888

http://www.super8.com

Super 8 Motels are clean, reliable, and affordable. This property has been renovated and updated recently. It is 20 miles to the airport from here.

#8

American Airpower Heritage Museum

American Airpower Heritage Museum is located at the **Commemorative Air Force** Headquarters in Midland, Texas. Most of us remember this as the home of the **Confederate Air Force.** The name has been changed I suppose to serve the needs of a world where political correctness has gotten very out of hand. This museum displays the largest collection of U.S. combat aircraft from the World War II era. Each is in flying condition.

I was particularly impressed by the **Aviation Art Gallery,** 10,000 square feet dedicated to preserving an impressive collection of World War II nose art. It may be the largest assemblage of this genre in the world.

This is a good museum, no doubt about it, and the planes that they keep here are absolutely pristine. There are many museums across the country that have birds from the same flock. Maybe not as many and maybe their facilities aren't quite as nice but the display is the same. So why would you come all the way to Midland to spend time at this one? While this is not the end of the earth, it can certainly be seen from here. Midland is smack in the middle of the desert known as west Texas. It is about half way between El Paso and Dallas which places it about 300 miles from either. You gotta' want to come here very badly, and the planes on display just aren't enough reason.

However, I recommend that you do come and that you do so often because this museum is dedicated to capturing and preserving the history of World War II. It does this better than all of the other museums focusing on that period combined. Two examples are striking: First, a few years ago they decided to tape record the stories of as many of the air warriors of WW II as they could. Today the **Oral History Program** has vaulted the memories of 2000. That is a good thing.

The lecture series will also draw you here. Throughout the year air warriors from World War II come here to relate their experiences. This opportunity should not be missed as it cannot continue for much longer. Most of these folks are in their early eighties. On March 19 and 20, 2004 the Museum will pay tribute to Col. Robert Morgan, pilot of the Memphis Belle. On the 19 the original Billy Wilder film will be shown. It is worth seeing just one more time, and I am certain that the accompanying dinner will be lavish. The next afternoon Col. Morgan will be here. He will be interviewed and you can and should be there to watch, listen, and learn from his stirring tale. Earlier in the year, on January 17 to be exact, Marine aviator Tom Danaher described World War II's last aerial.

Every place else you can see history; here you can shake its hand and say, *Thank you!*

Dates

Open daily
Closed Thanksgiving and Christmas Day

Hours

Monday–Saturday	9 a.m.–5 p.m.
Sunday and holidays	12 p.m.–5 p.m.

Admission Cost

Adults	$9.00
Teens (13–18)	$8.00
Children (6–12)	$6.00
Seniors	$8.00
Children (5 and under)	Free

Contact Information

American Airpower Heritage Museum
Midland International Airport
9600 Wright Drive
Midland, Texas 79711
(432) 567-3009
(432) 567-3047 (fax)
http://www.airpowermuseum.org/

G/A Airport Serving the Event

Midland, TX (Midland International—MAF)
Phone: (915) 560-2200
Tower: 118.7
ASOS: (915) 561-5135
Runways:

10–28	8302′ × 150′
4–22	4605′ × 75′
16R–34L	9501′ × 150′
16L–34R	4339′ × 100′

FBO Name/Phone Number

Avion Flight Centre, Inc.
(915) 563-2033

Ground Transportation

Rental car

Best Place to Stay $$

Fairfield Inn
2300 Faulkner Drive
Midland, TX 79705
(432) 570-7155

Best Place to Stay $$

Holiday Inn Hotel & Suites
4300 W. Wall
Midland, TX 79703
(432) 697-3181

#9

American Airpower Museum

American Airpower Museum in New York is the area's leading living aviation historical institution. "Living" because it has operational, flight-ready aircraft. Students and interested adults will witness these historic warbirds take flight when weather permits. **American Airpower Museum** at Republic Airport in New York is fast becoming an international destination for those seeking to experience the history symbolized by these aircraft as they take to the air. Republic Airport is a state-owned general aviation airport. It is only minutes from the Farmingdale, Long Island, railroad station.

The thrill begins the moment you arrive. The control tower was retrofitted to perfectly match the 1940s. One of the hangars appears as it did in Pearl Harbor on the eve of war. All were actually designed and built during World War II. Each is filled with operational warplanes from that era and staffed with volunteers in period flight gear. Vintage automobiles and trucks are sprinkled around to add to the ambiance. The goal of the **American Airpower Museum** is to bring to life powerful images from another era cost effectively. Believe me, they have achieved it!

The squadron of aircraft includes a B-17 Flying Fortress painted like the Memphis Belle, a B-25 Mitchell bomber, a P-40 Warhawk, painted like the Flying Tiger that made it famous. A Corsair and a Thunderbolt are here as are many others. All are flyable for this is the place *where history flies!*

The best reason to come here? This is truly the place where history flies and so can you. The museum is offering a D-Day Commemoration flight in its C-47. That's right, the military version of the DC-3. The plane that ferried tens of thousands of paratroopers across the ditch for the D-Day invasion. The flight experience is yours for a $250 tax-deductible gift to the museum. You will be issued "orders" instructing you to report to the museum complex at Republic Airport. Give the museum a call and they'll fill you in on all the details. Originally this was to be a one-time offering. I understand that more flights are being added. Don't miss the chance. A C-47 is like a DC-c but the ride is *very* different.

Dates

Open Thursday through Sunday

Hours

10:30 a.m. through 4:00 p.m.

Admission Cost

Adults	$9
Seniors	$6
Children	$4

Contact Information

American Airpower Museum
Republic Airport
Farmingdale, NY
(212) 843-8010
info@americanairpowermuseum.com
http://www.americanairpowermuseum.com

G/A Airport Serving the Event

Farmingdale, NY (Republic Field—FRG)
Phone: (631) 752-7707
Tower: 188.8
ASOS: (631) 752-8129
Runways:

1–19	5516′ × 150′
14–32	6827′ × 150′

FBO Name/Phone Number

Atlantic Long Island
(631) 752-9022
(631) 752-9099 (fax)
www.atlanticaviation.com

Ground Transportation

Rental car

Best Place to Stay $$$

Holiday Inn
215 Sunnyside Blvd.
Plainview, NY 11803
(516) 349-7400

Best Place to Stay $$$

Hampton Inn Long Island/Commack
680 Commack Road
Commack, NY 11725
(631) 462-5700

Best Aviation Attraction

#10

AOPA Expo

(*Courtesy AOPA.*)

The **Aircraft Owners and Pilots Association** with over 400,000 members certainly has reach with the upper end of the individual owner/pilot market. It is also the largest and strongest lobbying organization presenting the views of general aviation to a government that all too often forgets them. Each year AOPA brings the faithful together for a very upscale convention in one of the nation's "gotta' go there" destinations. The Long Beach Convention Center in Long Beach, California will host the 2004 event. With 500 industry exhibits, 60 aircraft in the static display, 80 seminar hours, and free general sessions, **AOPA Expo** is the premier general aviation event. AOPA Expo is a chance for aviation enthusiasts to learn more about general aviation.

More than 10,000 general aviation pilots, family members, friends, and others are expected to attend Expo 2004. It's a one-stop opportunity to check out new airplanes and equipment, learn the latest safety procedures, and enjoy the company of other aviation supports. Expo is not just for pilots. It is the ideal opportunity for anyone to get up close and personal with the largest segment of the aviation industry.

(*Courtesy AOPA.*)

AOPA Expo 2004 is open to the public and offers a wide spectrum of aviation events. Expo visitors can wander through the exhibition hall and the static aircraft display or attend one of the many seminars. The seminars are conveniently grouped by topic, including general interest, proficient pilot, all pilots, medical issues, safety seminars, and owner seminars.

The exhibit hall and product demonstrations open at 10 a.m. each day, closing at 6 p.m. the first two days and at 4 p.m. on the last day. Everything from aviation art to propellers will be shown in the convention hall. Serious shoppers will enjoy an opportunity to "try before they buy" with selected product demonstrations.

At least 60 general aviation aircraft, ranging from sporty two-seaters to corporate business jets will be on display. Most will be available for attendees to sit in and study, up-close and personal.

Complete and up-to-date information about AOPA Expo 2004 is available at the organization's Website (www.aopa.org).

Dates

October 21–23, 2004

Hours

Opens 10 a.m.
Closes 6 p.m.

Admission Cost

Exhibition hall and static display	$30 daily
Exhibition hall and static display and seminars	$45 daily

There are two- and three-day packages available.

Contact

AOPA Expo 2004
1 (888) GO2EXPO

G/A Airport Serving the Event

Long Beach, CA (Long Beach/Daugherty Field—LGB)
Phone: (562) 570-2678
Tower: 119.4
ASOS: (562) 424-0572
Runways: Several

FBO Name/Phone Number

Million Air Long Beach
(562) 981-2656
info.lgb@millionair.com

Ground Transportation

Rental car

Best Place to Stay $$$

Queen Mary
1126 Queens Highway
Long Beach, California 90802
Phone: (562) 435-3511
http://www.queenmary.com/

The Queen Mary is one of southern California's major tourists' attractions. What is general unknown is that this beached whale of a boat is also a sumptuous hotel. Many of the rooms are small, staterooms are like that, but you can book a very adequate suite for not much more than the nightly rate at other five-star hotels. Stay here and turn your AOPA Expo visit into a travel bonanza that will give you something to talk about for years to come.

Best Aviation Attraction

#11

Armchair Flyer

Airplanes are thrilling. Going to an air show or a museum is high on my list of things to do. Naturally, any chance to go flying is never passed up. I really like airplanes and I like being around people who think the same way. It turns out that I am not around aviation as much as I wish to be around it. That is true for most of you as well. At those times I often become an **Armchair Flyer.** I hope you will as well.

We can think about an activity more than we can do that activity. I am sure that no one will argue with that. I like to focus and direct my thoughts to allow these times to be truly productive. When it comes to aviation can you imagine a more productive evening than one spent with Charles Lindbergh? What would it be like to just sit and listen while he told you in his own words the real story of *The Spirit of St. Louis* and the flight that changed the world? It turns out that such an evening is available to you whenever you're ready.

Charles Lindbergh penned an autobiographical account of the flight. *The Spirit of St. Louis* by Charles A. Lindbergh, with an introduction by his son Reeve Lindbergh is one of two must reads for any **Armchair Flyer.** This is the book on which the movie *The Spirit of St. Louis* starring Jimmy Stewart was based. Jimmy Stewart playing Charles Lindbergh was casting perfection. They even look a little alike. Well, why not rent the movie and pass on the book? As you can well imagine, there is much in the book that isn't in the movie. Any autobiographical book gives you a pretty clear look inside the author: the way he uses language, the things he dwells on. What was important to Lindbergh about the airplane and the flight may not have been dramatic enough for the producer to include in the movie. Did you know that Charles Lindbergh claims in his book that there were mysterious beings on board the aircraft with him as he crossed the Atlantic? Were these delusions driven by sleep deprivation? Did he later believe that he saw these spirit-like beings floating in the aft fuselage or did he come to believe that he only imagined them? It is worth reading the book just to find out.

What was it about Lindbergh and his early years that drove him to take the risk of a flight across the Atlantic? Who was the dominant force in his early years? Lindbergh was a man driven as much by emotion as by science. He is a wonderful read.

The Wright brothers on the other hand were almost totally driven by a love of science. Every step they took that resulted in their monumental achievement of developing and controlling a machine in flight was preceded by study. Hypostasis, leading to theory, leading to formula, leading to experimentation, leading to law. The brothers personified the scientific method. Perhaps their real gift to mankind was not flight but structured thought. They were not like Edison. He was given to quick experimentation and many resulting failures until discovery. The Wrights met few failures when they arrived at the experimentation phase. They did their homework. Their achievement was not only powered flight. It was all the things they had to develop to be able to take that powered flight. One thing they needed was power. No aircraft-suitable engine existed; they designed and manufactured one. Amazing! That alone should earn them esteem in the eyes of all. Next they needed a way to translate that power to flight. They needed a propeller. It had to be perfect and certainly could not be purchased as none existed. For this too they should be held in high esteem.

The Wrights meticulously recorded all of their theories and experimental results and kept these documents. Later these papers would be gathered together, placed in a logical chronological order, and indexed painstakingly. The editing was done by Wilbur Wright. *The Papers of Wilbur & Orville Wright, Including the Chanute-Wright Papers,* edited by Wilbur Wright, is a pricey book. It can be found discounted at Amazon.com and other sources. You'll spend not one, not two, but several evenings reading and studying these papers. It is a big, thick, pithy book. Supplement it with *How We Invented the Airplane: An Illustrated History* by Orville Wright. This book is short and sweet and to the point. It is only 96 pages and contains several photos. With these two books you will understand the brothers and why they succeeded.

So, fellow **Armchair Flyers,** read these three books and you will have a pretty good understanding of the two most important events in the history of aviation. There is another book and another author I recommend to you, *North to the Orient* by Anne Morrow Lindbergh. Shortly after his journey to Paris, Charles Lindbergh met and married Anne Morrow. Together they set out on an epic journey in a float plane. It was their intention to fly to the Orient via Alaska and the Aleutians. Anne became an expert radio operator which in those days meant sending and receiving messages in Morse code. It also meant learning how to adjust the antenna, which was merely a trailed wire. Mrs. Lindbergh wrote the story of their flight in *North to the Orient.* It established her as a writer of great merit and was probably as important as the charts they made in opening up the air routes they had discovered. These are the same routes flown to this day.

I have one final favorite author, a Frenchman, Antoine De Saint-Exupery. He was a flyer in the golden years of aviation and flew the mail routes of northern African in the late years of the 1920s and the early years of the 1930s. Later he would defend France from the cockpit of a P-38. There he met his death in a flight over the Mediterranean. His tales of his early experiences are beautifully recorded in many volumes. The best in my mind was his masterful *Wind, Sand and Stars.* It is as wonderful a work of prose as you will ever find. Please know that while he flew the P-38 he took along pencil and paper and penned his highly revered book *The Little Prince.*

From Charles Lindbergh and the Wright brothers you will learn how man learned to fly and how he flew. From Anne Morrow Lindbergh and Antoine De Saint-Exupery you will be connected to the poetry of flight and the souls of the people that dare to duel with the wind and the stars.

Dates

Any quiet night

Hours

The lonely hours are best

Admission Cost

The price of a book

Contact Information

Your neighborhood bookstore or Amazon.com

Best Aviation Attraction

#12

B-24 Flight Experience

If it's good enough for Tom Hanks, it's good enough for you! Seeing airplanes is wonderful and watching airplanes fly is awesome but flying airplanes is the icing on any aficionado's cake. Flying is the stuff that memories are made of, I do believe.

My father was a B-24 flyer in World War II. Whenever I see one I become very nostalgic and wonder what it must have been like for him. A few years ago my brother found our Dad's World War II diary. I read it. That helped fill in the blanks, but I still wonder what it must have been like. This year I'm going to find out!

In 1979 **The Collings Foundation** was founded to support "living history" events involving transportation. Their original focus was on automobiles, but they quickly moved to aircraft. They have many planes which are displayed at their museum in Stowe, Massachusetts. Three of them fly across America each year on the **Wings of Freedom Tour,** a B-24, a B-17, and a newly added B-25. Maybe they've come to your town. A very special feature of the tour is the **Flight Experience** program.

B-24 Flights last approximately 30 minutes and require a $400 donation per person. A flight will not be made with less than six or more than eight passengers. B-25 Flights are approximately 25 minutes. The charge varies based on seating position. In the front (two seats behind pilot) a $350 donation per person is required. The back (four seats behind bomb bay) cost a little less, only a $300 donation per person is requested. A B-25 flight cannot be made with less than five or more than six passengers.

I'm going. Are you?

Contact Information

The Collings Foundation
(800) 568-8924
http://www.collingsfoundation.org

Best Aviation Attraction

#13

Big Bunny

Some years ago Hugh Heffner was busily trying to find ways to expand his ballooning empire and live life just a little larger. Not that things were bad for Hugh. Heck, he had the world famous Playboy Mansion in Chicago, a string of Playboy Clubs worldwide, the most successful magazine ever launched, hot and cold running women (Playmates), and a beautifully converted DC-9 he called the Big Bunny. He used this plane to move around the world and oversee the minions of Le Domain Hef. The idea came to him that if Playboy Clubs were a hit then a Playboy resort should be a smash and one with its own landing strip capable of handling the Big Bunny would be super OK. His thoughts turned to Lake Geneva, Wisconsin, not hard to do since he lived in nearby Chicago. Here he built his first and most impressive resort.

Time moved on and business turned sour. The clubs closed, the Big Bunny flew away, his daughter Christy took over the magazine, and he sold off the resort. The house that Hef built continued to slide downhill like a Wisconsin toboggan run for a few years. Then it was purchased by a new group that threw big bucks into renovating the resort to a level that would make Hugh blush! This place put the *p* in plush. Naturally it sports the coveted AAA four-diamond award. It is a place to be pampered. Ski here in the winter, golf the rest of year, and spa anytime you feel the need.

You pilots can land right here. The resort has its own airport, a nice one at that. The 4100-foot runway offers complimentary overnight tie-down and free shuttle service between the airport and the main resort. Land here for brunch, a day at the spa, a round of golf, or an overnight romantic husband and wife midweek getaway.

The **Grand Geneva Resort and Spa** is in my opinion the best fly-in resort in the Mid-West. Stop in and see if you don't agree.

Contact Information

Grand Geneva Resort
7036 Grand Geneva Way
Lake Geneva, WI 53147
(800) 558-3417
(262) 248-8811
grandgeneva.info@grandgeneva.com
http://www.grandgeneva.com/

G/A Airport Serving the Event

Lake Geneva, WI (Grand Geneva Resort—C02)
Phone: (262) 248-8811
CTAF-Unicom: 122.8
Runway: 5–23 4100′ × 100′

FBO Name/Phone Number

Grand Geneva Resort Airport
(262) 248-8811

Ground Transportation

None needed

Best Place to Stay $$$$$

Grand Geneva Resort
7036 Grand Geneva Way
Lake Geneva, WI 53147
(800) 558-3417
(262) 248-8811
grandgeneva.info@grandgeneva.com
http://www.grandgeneva.com/

#14

Boneyard—Aerospace Maintenance and Regeneration Center

The **Aerospace Maintenance and Regeneration Center (AMARC)** or simply the **Boneyard** is the largest collection of aircraft anywhere in the world. There are over 4000 planes parked here, in row after endless row. Most are from the Vietnam era but there are also F-16s and A-10 "Warthogs," from the Gulf War. None of these aircraft are flyable. The U.S. Air Force *mothballs* planes here until they need them again. Some are reactivated for the Air Force's use, some are sold to foreign governments, and some are parted out. Amazingly enough, some are turned into pilotless drones and used for missile target practice.

This is something that you really do want to see. Weekday tours of **AMARC** are offered by the Arizona Aerospace Foundation, which operates the **Pima Air and Space Museum** and **Titan Missile Museum.** For more information on tour schedules contact the **Pima Air and Space Museum** at (520) 574-0462.

The tours have proven popular, so reservations are strongly recommended to guarantee seating. Seven days notice is preferable when making a reservation.

Check in at the **Pima Air and Space Museum** gift shop no later than 30 minutes prior to tour departure time or your reservation may be subject to cancellation. The tour bus boards at the Museum entrance. Due to increased security concerns, one small carry-on item is allowed per passenger. Carry-on items may include a small camera, purse, or fanny/belt pack. Photo identification is required for everyone 16 years and over.

Dates

Tours are given Monday through Friday except on New Year's Day, Martin Luther King Jr.'s Birthday, Presidents' Day, Memorial Day, Independence Day, Labor Day, Columbus Day, Veterans Day, Thanksgiving Day, Day following Thanksgiving, Christmas Eve, Christmas Day.

Hours

May 1–September 30	11:30 a.m. & 1:30 p.m.
October 1–May 31	10:00 a.m., 12:00 a.m., and 2:00 p.m.

Admission Cost

Adults (ages 13 and up)	$6.00

Groups of 20 or more	$5.50
Children 12 and under	$3.00

Contact Information

Advance reservations are *strongly* recommended to guarantee seating. For more information on tour schedules contact:

John Lundquist
Pima Air and Space Museum
6000 East Valencia Road
Tucson, AZ 85706
(520) 574-0462
(520) 574-9238
http://www.pimaair.org/amarc.htm
http://www.dm.af.mil/AMARC/

G/A Airport Serving the Museum

Tucson, AZ (Tucson International—TUS)
Website: www.tucsonairport.org
Phone: (520) 573-8100
Tower: 118.3
ASOS: (520) 889-7236
Runways:

11R–29L	11,000′ × 150′
11L–29R	8400′ × 75′
3–21	7000′ × 150′

FBO Name/Phone Number

Tucson Executive Terminal
(520) 573-8128

Ground Transportation

Enterprise Rent A Car
(800) Rent A Car

Best Place to Stay $$$$$

The Westin La Paloma Resort and Spa
3800 East Sunrise Drive
Tucson, AZ 85718
(520) 742-6000
http://www.westinlapalomaresort.com/

It is approximately 20 miles from the museum and well worth the trip. North of Tucson in the lush Sonora Desert, nestled among the foothills of the Santa Catalina Mountains. The Westin La Paloma Resort and Spa offers luxurious guest rooms, a myriad of recreational activities, and the superb service expected of a AAA four-diamond-award resort. It's easy to see why it was named to the 2002 *Condé Nast Traveler Gold List.*

Best Place to Stay $

Best Western Inn at the Airport
7060 S. Tucson Boulevard
Tucson, AZ 85706
(520) 746-0271

Conveniently located adjacent to Tucson International Airport is the Best Western Inn at the Airport. This is a great choice for four reasons:

- You can't get any closer to the airport and they run a free shuttle.
- It is only 5 miles to the museum.
- This property has a pool and a putting green.
- The value is exceptional! You should be able to get a room for under $50.

#15

Buy an Airplane

It starts with a dream and the knowledge that if you can dream it you can do it. Me own an airplane? Of course you can. First things first; if you're not a licensed pilot but want to become one, this may be the best time for you to buy an airplane. To meet the FAA requirements to become a private pilot you must have 40 hours of flying time, much of it solo without an instructor. Most people will not be able to become proficient enough to pass the check ride with only 40 hours. The average new pilot will have logged more than 50 hours by the time he earns his private pilot license with a single engine land rating. Most trainees will solo in less than 12 hours. At that point, much of the training will be done without your instructor on board.

A brand new Cessna 172SP can be rented at any Cessna Pilot Center for $115 per hour wet. An older 172 with more hours on it will go for $90 per hour also wet. The word *wet* in this case simply means that you'll rent it with fuel and oil included. Your actual cost for using the aircraft is the stated hourly rate with nothing additional. Some flight schools will quote the rental charge dry (without fuel and oil) but that is rare. The instructor will cost you another $20 per hour.

By the time you receive your license you will have spent somewhere around $5300 on aircraft rental. The Cessna 172 burns about 10 gallons of 100 low-lead fuel per hour at an average cost of $2.50 per gallon. The fuel cost will vary widely based on where you buy it. Some owner/pilots will get an auto fuel waiver for their ship allowing them to burn auto fuel in it. In which case the fuel cost can drop all the way down to $1.40 per gallon. For the moment, using the 100 low-lead figure, it comes to $25 per hour or a total of $1250 for your flight training. Subtract that from the previously calculate $5300 rental cost to see what you're paying for just the aircraft. If arithmetic still works that's $4050. A brand new VFR equipped Cessna 172 will set you back about $155,000. That is too much to pay for an airplane that you may not want after your training is complete.

A great plan is to buy a used Cessna 172. You should easily be able to find a good VFR equipped 172 with a midtime engine for $40,000 or less. That $4050 you are about to pay in rental charges will take care of 10 percent of this ship's purchase price. The best news is that used aircraft prices tend to go up not down.

Here's a good idea for you. Most flight instructors are young men and women who aspire to be airline pilots. They work as flight instructors because it's a way for them to put flight time in their logbook. This helps them get selected for that coveted airline driver's job. They'll do anything for flight time. So now that you're an aircraft owner work a deal, trade an hour's free use of your ship (dry) for an hour's worth of flight instruction. Your total out of pocket cost for getting your license will be the $1250 fuel bill. If you sell your aircraft at the end of the flight training experience you will get about what you paid for it maybe a little bit more.

If you already have your license, know that this is the best time in history to buy an airplane, new or used. There are terrific tax implications available for purchasing a new plane. Call a dealer that handles new aircraft and have the salesperson fill you in.

Whether you buy new or used, interest rates haven't been lower in about two generations. If you happen to be a homeowner you can turn those airplane payments into tax exemptions by refinancing your home and pulling out equity dollars to buy the plane rather than taking out a loan for the ship directly. For whatever reason, our government allows tax deductions for the interest expense on primary home mortgages and not for automobile, boat, or aircraft loans. So work the system and let the IRS help pay for your plane.

Personally I would never buy a new airplane. Its value is going to fall while the value of used aircraft has historically risen. Those are the facts.

Used aircraft prices are largely driven by the time on the engine and the avionics in the panel. Secondary issues are the maintenance history of the plane and engine, the condition of the paint and interior, and the date of its last annual inspection. A good place to start looking for your airplane is on the Internet. The two best places offering used aircraft listings are www.aso.com and www.tradeaplane.com.

Operating costs of an airplane are more than fuel and oil. It needs to live somewhere. Check with local airports before you buy the airplane and get a good idea about where your machine is going to be and whether it is going to live inside a hangar or on a tie-down. Hangars keep airplanes and their paint jobs nice and fresh, but if the cost difference is substantial you may find that a tie-down is a better alternative. The cost difference in my area is so great (there aren't enough hangars) that leaving an airplane outside and buying a new paint job every 4 years is a moneymaker. Don't forget about the cost of maintenance. It will be slightly more than the annual inspection. Look hard to find an honest A&P with an IA to take care of your ship and do the annual. It will make a world of difference in the quality of your flying experience and the health of your wallet.

If you love airplanes and aviation sooner or later you gotta' take the plunge. Owning your ship is a much better experience than renting someone else's. The funny thing is, it's also cheaper!

Let your wallet be your guide and find a ship that you can afford. There are good grass-strip airplanes out there that can be yours for as little as $15,000. Now they won't carry a heavy load or get you to grandma's house at Christmas but they can be yours and they are obtainable.

If your wallet is a little thicker do consider new airplanes but not a new Cessna, Piper, or Beechcraft. There is nothing substantially different between a new Piper Archer and a used Piper Archer. It is principally the same airplane they have been building since the Cherokee 180 days. That was about 40 years ago. So buy a used one and save $150,000. Cirrus, Diamond, and Lancair are exceptions. Their planes are as new as tomorrow and employ totally new technology everywhere. Consider the soon to be delivered DA42 Twin Star from Diamond. It is a twin-engine aircraft using diesel engines. The benefits are great. It will burn jet A which is much less expensive than 100 low lead. Its fuel burn is only 10 gallons per hour, for *both* engines! This economy bullet will cruise at over 200 mph with a range of well over 1200 miles. It is made from space-age composite materials proven to not corrode. Look at the *new* airplanes. If you're going to keep it for a while buy as big as you can afford and don't overlook all the terrific personal jets that are coming down the chute right now!

Best Aviation Attraction

#16

Canada Aviation Museum

Canada is one of two countries in the world that has weaved private aviation into the fabric of everyday life. The other is Australia. The colonization of this vast land could not have occurred without private aviation. Much of what makes Canada livable today is dependent on that rare breed we call *bush pilots.* What we often forget is that thousands of ordinary Canadian citizens fly almost daily not as a profession but as part of their life. Canada is about civilian aviation more than any other country in the world! This museum tells the tale of Canadian aviation. It is a must see!

The **Canada Aviation Museum** at Uplands Airport in Ottawa was originally one of three major collections owned by the Canadian government. It was immediately recognized as having the most extensive aviation collection in Canada and ranked among the best in the world.

In 1964, the three collections were combined under the name **National Aeronautical Collection,** and relocated to Ottawa's historic Rockcliffe Airport. The arrangement gives visitors a rich perspective of the development and use of aircraft in Canada over the years.

In 1967 the National Museum of Science and Technology assumed administration of the National Aeronautical Collection. It continued to acquire both military and civil aircraft important to Canada's aviation history. Today it includes more than 120 aircraft. In 2000, the **National Aviation Museum** was officially renamed the **Canada Aviation Museum** and remains a component of the **Museum of the Canada Science and Technology Corporation.**

The **Canada Aviation Museum's** collection policy is to illustrate the development of the flying machine in peace and war from the pioneering period to the present time. The collection naturally focuses on Canadian achievements. However, aircraft from many nations are represented.

Dates

Open seven days a week during the summer
Closed on Monday and Tuesday during the winter
Closed on Christmas Day

Hours

Summer 9:00 a.m.–5:00 p.m.

Winter 9:00 a.m.–5:00 p.m.

Admission Cost

Adults	$6
Students	$5
Seniors	$4
Children (6–15)	$3
Children (under 6)	Free
Families (1 or 2 adults with children)	$13
Daily (4:00 p.m. to 5:00 p.m.)	Free

Contact Information

Canada Aviation Museum
11 Aviation Parkway
Ottawa, Ontario
K1K 4R3
(613) 993-2010
(800) 463-2038
aviation@technomuses.ca
http://www.nmstc.ca/nam/Eng/english_home.html

G/A Airport Serving the Event

Ottawa, Ontario (Ottawa-Rockcliffe Airport—CYRO)

FBO Name/Phone Number:

When you arrive in Ottawa, contact the **Rockcliffe Flying Club** at radio frequency 123.5.

They will notify the museum of your arrival. After landing, taxi to the museum and park at the north east corner of the museum building. A member of the museum's visitor services staff will greet you and escort you into the museum. It's that easy!

Ground Transportation

Rental car

Best Place to Stay $$$$

Fairmont Château Laurier
1 Rideau St
Ottawa, ON K1N 8S7
Canada
(613) 241-1414
http://www.fairmonthotels.com/

This is the finest example of period architecture in Canada. Take the time to stay here. Winter rates are often lower than $150 per night.

Best Aviation Attraction

#17

Canadian International Air Show

(*Courtesy the United States Navy.*)

The **Canadian International Air Show** takes place in Toronto. It is held annually over the Canadian Labor Day weekend which will occur on the first weekend of September in 2004. The three-day event will begin on Saturday, September 4, and end on Monday, September 6, 2004. I particularly like this show because, like the shows in

Chicago and Cleveland, it takes place downtown along the lake shore. This affords really good viewing. Where are the best places? Not in the reviewing stands. The best places are downtown in the high-rise buildings. The skyline is constantly changing so plan to do some last-minute research. This may radically affect your choice of a hotel.

Citizens of the United States like airplanes, Canadians love them. Their infatuation causes a huge turnout. Thank goodness the crowd is spread over miles of shoreline. Canadians are also fiercely loyal to The Snowbirds. This is their flight demonstration team. The team's pilots are national heroes and are well known throughout the country.

In Toronto, and nowhere else, there is a very special twilight show. Not fireworks, aircraft! It kicks off on Friday after the regular show has ended. Normally it starts at 6:45 p.m. and continues until about 8:00 p.m. The Snowbirds are the featured act but watch for contemporary fighters as well. Both the U.S. and Canadian Air Forces will fly the latest equipment including F-16s, F-15s, F-18s, and F-14s. Twilight is long and lovely this far north and the planes are shown off in a way that you seldom see.

Dates

Begins September 4, 2004
Closes September 6, 2004

Admission Cost

General Admission	$10.00
Seniors	$7.00
Children	$7.00
Babes in arms	Free

There are no tickets required to watch the air show. The admission charge is for watching from Ontario Place or the CNE.

Any location giving a view of the section of Lake Ontario just west of Ontario Place is great for viewing the air show.

Contact Information

Canadian International Air Show
CNE Press Building, First Floor
Exhibition Place, Toronto
M6K 3C3
(416) 263-3650
(416) 263-3838 (fax)
admin@cias.org
http://www.cias.org/

G/A Airport Serving the Event

Toronto, Ontario, Canada (Toronto City Centre—CYTZ)
60 Harbour Street
Toronto, Ontario, Canada M5J 1B7
(416) 203-6942
(416) 203-6741 (fax)
tcca@torontoport.com
Tower: 118.2
ATIS: 133.6
Runways:

08–26	4000′ × 150′
06–24	3000′ × 150′
15–33	2780′ × 150′

Be warned! There is a minimum landing fee of $10.75 Cdn for Piston Aircraft.

Ground Transportation

Rental cars

Best Place to Stay $$$

Hilton Toronto Airport
5875 Airport Rd.
Mississauga, ON L4V 1N1
(905) 677-9900
(905) 677-5073 (fax)

Best Place to Stay $

Days Hotel & Conference Centre—Toronto Airport
6257 Airport Rd.
Mississauga, ON L4V 1E4
(905) 678-1400
(905) 678-9130 (fax)

#18

Castle Air Museum

Castle Air Force Base in Atwater, California, became Castle Airport in 1994. At the same time a group of local airplane enthusiasts formed the **Castle Air Museum Foundation, Inc.,** a nonprofit organization. Their purpose was to assume custody of the collection of aircraft already associated with the base.

Castle Air Museum is continuing to add to its collection of historic World War II, Korean War, and Vietnam Conflict aircraft. Currently the displays include almost 50 aircraft. For me, the trip is made by the B-24 Liberator. It reminds me of the one my father served on in World War II. My other favorites are here as well: a SR-71 Blackbird (the fastest aircraft ever manufactured), a De Havilland U-6A Beaver (the military version), and a RB-36 Peacemaker.

One of the best reasons to visit this museum is the overnight accommodations you can enjoy. Gemini Flight Services, the airport's FBO, operates the base's TLQs (transient living quarters). These are the exact building, rooms, and furniture that visiting air crews used when this was Castle Air Force Base. Relax in the TLQ park and enjoy the view of outside displayed aircraft at **Castle Air Museum.**

Dates

Open daily
Closed Tuesday
Closed New Year's Day, Easter, Thanksgiving, Christmas Eve, and Christmas

Hours

May 1 through September 30 9:00 a.m. to 5:00 p.m.
October 1 through April 30 10:00 a.m. to 4:00 p.m.

Admission Cost

Adults	$7.00
Seniors (60 and up)	$5.00
Youths (8–17)	$5.00
Children (7 and under)	Free

Open Cockpit Day

Adults	$10.00
Seniors (60 and up)	$8.00
Youths (12–17)	$8.00
Youths (8–11)	$6.00
Children (7 and under)	Free

Contact Information

Castle Air Museum
5050 Santa Fe Drive
Atwater, CA 95301
(209) 723-2178
(209) 723-0323 (fax)
cam@elite.net
http://www.elite.net/castle-air/

G/A Airport Serving the Event

Atwater, CA (Castle—MER)
Phone: (209) 384-7325
CTAF-Unicom: 123.0
Runway: 13–31 11,802′ × 300′

You can land on the fourth longest runway on the West Coast. Taxi to Gemini Flight Support, check into the TLQ and walk to the museum. This is one of general aviation's best fly-to experiences. If you are a pilot and you're visiting California without your ship, rent one somewhere close and fly to this museum; it is really worth the trip!

FBO Name/Phone Number

Gemini Flight Support
(209) 725-1455
gemini@elite.net

Ground Transportation

Rental Car

Best Place to Stay $

Gemini Billeting
Castle Airport
(209) 725-1455
http://geminiflightsupport.com
gemini@elite.net

It doesn't get any better than this ever. You'll be put up in the old Castle Air Force Base transient living quarters. There are singles and doubles available. The doubles use bunk beds. You may even cook in these rooms if you like. The best part is that you can walk to the Museum from here.

#19

Cayman Caravan

International Aviation Week in the Cayman Islands consists of seminars, an airshow, and lots of aviation-related activities, including aerobatic demos, skydiving, and static displays, all hosted at **Owen Roberts Airport on Grand Cayman Island.** It has been going strong since its inception in 1986.

The best beach in the Caribbean, scuba diving, and terrific resorts make this the perfect outing for private pilots and their families, except for two things. First, more than 50 percent of the 330-mile flight from the closest U.S. airport in Key West, Florida, is over open ocean. Second, 70 miles of the trip is over Cuba. Both are daunting to weekend pilots.

There are two great options. It is only a 70-minute flight by commercial airliner from Miami. This first option is unthinkable to adventure-primed private pilots. The second option is using safety in numbers and leaving the details to professionals who have done it for a dozen years. The little details include getting overflight permission from the Cuban government. The safety in numbers comes from being part of 100-plus private airplanes that gather in Key West for the flight. All depart in groups of four for the 330-mile flight. The departures are carefully planned to place aircraft of similar performance together. They fly in trail, not in formation, and they must be placed a safe and legal separation distance from each other. This planned trip to **Cayman International Aviation Week** is called the **Cayman Caravan.**

The air show brings together military and private pilots. They will be making low-flying aerobatic passes along Seven Mile Beach. Aircraft featured in the show usually include: F14s, F16s, F18s and private aerobatic planes. There will be a static display at the General Aviation Terminal. It is good to lay outstretched on Seven Mile Beach, mud slide in one hand and cheeseburger in the other while an F-16 performs an aerial ballet above your splashing children.

Life is short— fly fast!

Dates

June 14–June 21, 2004

Admission Cost

The registration fee is $449 per aircraft ($499 after Tuesday, May 18, 2004). The fee covers the following services:

Trip kit (with all the required paperwork filled out and waiting for your signature)

Customs expediting (U.S. and Grand Cayman)

Cuban overflight permit
All charts and plates plus a detailed pilot flying guide
Prefiled flight plan services
Discounts on hotels and avgas
Reduced or eliminated customs and immigration fees
Water survival, ditching, and thunderstorm flying seminars in Key West
Transportation in Key West
Preflight briefing
Welcoming reception in Grand Cayman
Aviation Week mixer
Culinary caravan

Contact Information

Cayman Caravan
(850) 872-2495
http://www.cayman-caravan.com/
registrar@cayman-caravan.com

G/A Airport Serving the Event:

Georgetown, Grand Cayman Island (Owen Roberts Airport)

Best Aviation Attraction

#20

Cessna Pilot Centers—Take to the Sky!

(©*Renée Purner.*)

This is the greatest aviation attraction of them all and you're the star of the show. After you've visited all the museums, seen 47 airshows, and sucked up the history of Kitty Hawk, the moment comes when you must get into the air, personally not vicariously.

With 300-plus Cessna Pilot Centers (CPCs) throughout the world, one is near you. Each stands ready to quench your thirst for a true aviation adventure. Forty nine dollars buys you a Discovery Flight Coupon. You'll learn how airplanes fly, conduct a preflight, and actually take the controls, all under the guidance of a top-quality flight instructor.

Hundreds of aviation companies litter the bankruptcy landscape. The one Clyde Cessna founded 75 years ago is still in business and thriving. Over half the general aviation aircraft flying today are Cessnas. More than 184,000 have been produced. The Cessna 172 *Skyhawk* is the most popular single-engine, personal-use aircraft ever produced. While fun to fly, the *Skyhawk* is forgiving, and efficient. It's the mainstay of every Cessna Pilot Center.

All Cessna Pilot Centers are associated with The Cessna Aircraft Company. They maintain the vision that has helped Cessna remain successful. Cessna Pilot Centers meet higher standards than the typical flight school in order to be eligible to operate under the Cessna logo. They must be clean, neat, and provide superior customer service and flight training. Each must make available for flight training, fully insured, *new* Cessna 172 aircraft that are currently under factory warranty. A Cessna Pilot Center is a specially selected flight school that employs only well-trained, professional instructors.

No matter which CPC you go to you'll need a Discovery Flight Coupon. You can get them online at http://learntofly.com/howto/discovery.chtml. Submit your information and print out the Discovery Flight Coupon. But there is an easier way! Visit your local Cessna Pilot Center and request the coupon, or contact Cessna Pilot Centers at (316) 517-3530.

Don't pass up this opportunity. It is the best chance you'll ever have to take to the sky with a professional flight instructor in a near new aircraft, under the watchful eye of a major aerospace company. If you get hooked, it is never too late to get a pilot's license. My friend from Wharton, Texas, Dick George, got his when he was almost 80 years old. Today he owns a *Skyhawk* and flies almost everyday!

Dates

Anytime you chose

Hours

Daylight is best but the skies are gorgeous at night!

Admission Cost

$49.00

Contact Information

Cessna Pilot Center
(877) 359-2373
(316) 517-6056
http://learntofly.com/

G/A Airport serving the event:

Two hundred plus worldwide

Best Aviation Attraction

#21

Charles de Gaulle Airport

As travelers we are very affected by the quality of the commercial aviation system. The airliners are uniform regardless of where you board or disembark but there is great difference in quality from airport to airport. I have spent much of my life getting on and off airplanes in many places around the world. It has left me with a feeling of dread when I know that a flight I am taking must terminate or connect through some airports. The worst on my list is Dallas Fort Worth. I have never experienced an on-time departure or arrival there, not one. In that I based out of Texas for more than a few years that is pretty sad. I also noticed that it is the least passenger-friendly airport I have ever happened upon. If you are changing planes at DFW, get your hiking shoes on, you are normally in for a pretty long walk even if you stay on the same airline. If you change airlines, you are necessarily relegated to the airport train which is best described as cumbersome and slow.

The contrast is Charles de Gaulle, the larger of Paris's two commercial airports. It handles more traffic but manages to do it with style and grace. Terminal 1, with which I am most familiar, and the other two are built with the passenger in mind. The check-in counters are only a few steps from the curbside drop-off areas. Once checked in you proceed to the inner area of this circular building where you are whisked to the shopping, restaurant, and gate area by a magnificent escalator which is suspended inside a Plexiglas tube. It is eye-catching and very efficient.

The shopping arcade is filled with the best European brands offered at duty-free prices. It is a good place to pick up gifts. The food offered in the various restaurants and bistros is typical French fare—real darn good! The walk from the shopping arcade to the gates is hardly noticeable.

If leaving from CDG is a breeze then arriving can only be described as a snap. Walk off the plane, pick up your luggage and a no-charge luggage cart for the promenade through customs. When you clear, you are within a few feet of curbside pickup from which you may select a taxi (don't), a limousine, an Air France bus, or an Orly bus to take you to central Paris. Certainly you could rent a car but I never do; driving in Paris is a blood sport and I never learned the art of it.

There is a better way to get to Paris; once you use it you will wish that all cities with remote airports had one this good. It is a TGV train station. From here you may take one of the high-speed trains to other regions of France or other European countries. To get to the center of Paris take the RER train. You will be there very quickly and very happily. The first stop is Gare du Nord.

By the way, Terminal 1 is the oldest and least efficient design of the two commercial terminals at CDG. The buildings that go together to make up the much newer and more modern Terminal 2 will knock your socks off. Aviation is about aircraft, airports, and air terminals. It is important to take a trip and become familiar with what they could all be like if our money was just spent a little more wisely.

Dates

Daily

Hours

24 hours a day

Contact Information

Paris Charles de Gaulle Airport
BP 20101
95711
Roissy/Charles de Gaulle
(0)1 4862 2280 or 1212
(0)1 4862 0752 or 5802 (fax)
info@adp.fr
www.adp.fr

Best Aviation Attraction

#22

Charter a Jet

You may have noticed that very few heavy hitters are flying the airlines any more even in the first-class cabin. They're still flying but just not in crowded, inconvenient commercial aviation. They've moved on to something much better. You're probably aware of the reasons they've moved on. Private jets fly on the traveler's schedule not their own. They go when you're ready not the other way round. That is a huge advantage. They are totally secure. You know everyone on board and in the flight cockpit. You are totally aware of the contents that went into the baggage compartment. They are secure but not inconvenient. You board them without going through a metal detector or being asked to remove any article of clothing. There are over 6000 airports in the United States, fewer than 500 are served by commercial airlines. That means that when the plane lands you can still have a 2-hour road trip in front of you in a rental car. Plus the routing to your intended destination is direct, rather than the hub-and-spoke system. You know the problem—to get from Birmingham to Huntsville on an airliner you have to first go to Atlanta. A 1-hour trip turns into 4 hours if you're lucky and make your connection, 8 hours if you miss it!

The solution of course is a private jet with its own flight crew. Most of us can't afford a private jet, new or used! Nor can we afford a share of a fractional jet program. Many of us can afford the usage fee to charter a jet at least some of the time. Look, I own my own single-engine aircraft and I use it for short trips of more than 100 miles and less than 400 miles; for those it is terrific. Nothing can give me and my family of three better bang for our buck. If it is 100 miles away or less we take the car. Between 100 and 400 miles and we're in the Cherokee. Distances of more than 400 miles have been the problem. I hate traveling on the airlines. It is incredibly inconvenient and very uncomfortable. For years both my wife and I traveled on business almost constantly so we know about the deteriorating quality of airline travel. She is actually a million-mile member with American Airlines.

Every fourth trip, which is about what our budget can handle, we charter. It isn't that much more expensive than a handful of first-class tickets on an airline. Let's say you're going from Orlando to New Orleans. That's about 2 hours in the air if you get a direct flight. There is no direct airline schedule. All flights connect somewhere; the most ridiculous connections being Houston and Miami. The travel time for this first-class 2-hour trip is almost 7 hours! The cost is $580 one way. Let's say we charter a six-passenger light jet for the same trip and fill all six seats. The cost is $1800 per hour for two hours or $3600. That's a lot of money for sure, but it is only $120 more than flying commercial and we get there 4½ hours sooner. Obviously empty seats in a charter are expensive.

So here's the challenge. Just once in your life let it rip! Get a group together and charter a jet. Go to a football game or the Indianapolis 500. Do something really fun that you'll always remember. Take your camera and make some memories!

Below are the names of two companies that can get you on your way. One is dedicated to backhauls—jets that have been rented for one-way trips and are flying back empty. You can get some real bargains. The other is a high-quality, make-it-happen, do-it-right, charge-a-fair-price kinda' company. Start your hunt with these two and use the Internet to hunt up others. This is an emerging industry and change is bound to come, particularly in pricing. It's going down believe me!

Contact Information

Private Business Jets
Norwell, Massachusetts
(800) 641-5387
http://www.flyprivate.com/
fly@flyprivate.com

Executive Air Charter (backhauls specialists)
Miami Springs, FL
(866) LEGSEEK
www.LEGSEEK.com
info@aircharternetwork.com

#23
Chicago Air & Water Show

The **Chicago Air & Water Show** is the oldest and largest *free* admission air and water exhibition of its kind in the United States, featuring civilian and U.S. military aircraft and watercraft.

The event kicks off each day at 9 a.m. with the Shell Extreme Watershow, featuring the Great Lakes Pro Freestyle Tour, the Bottoms-Up Boat Jersey Skiff Shoot-Out, and the Munson Ski & Inboard Water Sports "Ski Show Team."

The U.S. Air Force Thunderbirds return as the Chicago Air & Water Show's headlining act, performing precision aerial maneuvers that demonstrate the capabilities of Air Force high-performance aircraft. The squadron exhibits the professional qualities that the Air Force develops in the people who fly, maintain, and support these aircraft. Other air show performers include the Northern Lights Aerobatic Team, AeroShell Aerobatic Team, Lima Lima Flight Team, Sean D. Tucker & Team Oracle, Red Baron Stearman Squadron, and announcer Herb Hunter, who will return as the "Voice of the Chicago Air and Water Show" (please note: performers subject to change).

Dates

August 21–22, 2004

Hours

Water Show	9 a.m. both days
Air Show	11 a.m. until 4 p.m. both days
Twilight Air Show	6:45 p.m. until 8:30 p.m., focused at Navy Pier, *August 17 only*
Fireworks	10:15 p.m. at Navy Pier, *August 17 only*

Location

Fullerton Avenue to Oak Street, with North Avenue Beach as the focal point

Crowd Size

2 million

Admission Cost

Free

Contact

The Mayor's Office—Special Events (312) 744-3315
moseinquiry@cityofchicago.org

Website

http://www.egov.cityofchicago.org

Featured Aircraft

A-10 Thunderbolt, UH-60 Blackhawk, AH-64 Apache Helicopters, C130, C141, KC135, KC10, USN F-18 Hornet Demonstration, USN F-14 Tomcat Demonstration, USAF 920 Rescue Group, USAF T-38 Demonstration, USAF F-15 Eagle, IL Air National Guard F-16, USAF B1-B, USAF B2 Stealth Bomber and the USAF F117 Stealth Fighter

Featured Performers

The U.S. Navy Blue Angels
The U.S. Army Parachute Team Golden Knights
Aeroshell Aerobatic Team
Lima Lima Flight Team
Sean D. Tucker & Team Oracle
Red Baron Stearman Squadron
Ian Groom presented by FedEx Express
Red Bull Mig Magic

Visitor Hints

On the 21 and 22, the show is best seen from Lincoln Park. Set up your picnic blanket, chairs, and ice chest on the grassy meadow just west of Lakeshore Drive and south of Fullerton. If you're in the mood for shoulder-to-shoulder fans, stand on the sand at the North Avenue Beach. Get there early!!

The airshow acts practice for 2 days prior to the actual show. The 20 is a full rehearsal. Arrive a day early and see the show without the crowds. For a real thrill, have lunch at the top of the John Hancock building and watch the Blues crisscross the Michigan Avenue canyon!

If you're on a big-buck budget, stay at the Drake Hotel. Rent any room that faces north. You'll look straight down the shore line and see all the action. Book early and stay above the fifth floor.

G/A Airports Serving the Event

Palwaukee Municipal Airport (PWK)
Chicago Midway International Airport (MDW)

FBO Name/Phone Number

North American Jet (PWK)
(888) 359-6244
Signature Flight Support (PWK)
(847) 537-1200
Million Air (MDW)
(773) 284-2867
Atlantic Aviation (MDW)
(773) 582-5720
Signature Flight Support (MDW)
(773) 767-4400

Ground Transportation

From Palwaukee Municipal Airport (PWK)

Signature Flight Support (847) 537-1200

From Chicago Midway International Airport

Alamo (800) 327-9633
Avis (800) 831-2847

Budget	(800) 517-0700
Dollar	(800) 800-4000
Enterprise	(800) 566-9249
Hertz	(800) 654-3131
National	(800) 227-7368
Thrifty	(800) 527-7075

Best Place to Stay $$$$$

The Drake Hotel
140 East Walton Place
Chicago, IL 60611
(312) 787-1431
www.thedrakehotel.com

The Drake Hotel is a Chicago landmark and has been a symbol of white-glove elegance for over 80 years. The pride of Chicago, the hotel has been the first choice of celebrities and heads of state since its opening in 1920. Just 30 minutes' drive from Chicago's airports, O'Hare and Midway, The Drake Hotel is perfectly located in the heart of the Gold Coast, overlooking Lake Michigan, across from Oak Street Beach, and right on the Magnificent Mile in downtown Chicago. The Drake Hotel puts its guests in the center of Chicago's exciting shopping, nightlife, culture, and dining experiences.

Since 1920, The Drake Hotel has combined tradition with elegance and style to accommodate all the needs of its hotel guests. The hotel's 537 elegant rooms, including 74 luxurious suites, many with Chicago views, offer an array of modern amenities that include an executive desk with modem hookups, on-demand video, telephones with voicemail and dataport, and bathrobes.

Best Places to Stay $

Days Inn
1816 N. Clark Street
Chicago, IL 60614
(312) 664-3040
www.daysinn.com

Located in the safe residential neighborhood of Lincoln Park, the Days Inn Gold Coast boasts newly renovated rooms with incredible views of Lake Michigan and the Chicago skyline.

- Jog or bike through Lincoln Park
- Spend an afternoon sunning at North Avenue or Oak Street beaches
- Feed the animals at the Lincoln Park Zoo
- Stroll through Old Town and the Gold Coast where you will find beautiful homes, shops, and fine dining
- Don't forget to visit the John Hancock Building, Water Tower Place, and Michigan Avenue's Magnificent Mile

Comfort Inn
15 E. Ohio Street
Chicago, IL 60611
(312) 894-0900
www.comfortinn.com

Along Michigan Avenue. Shopping, dining, entertainment, historical building. Modern amenities; suites and standard rooms. Complimentary breakfast, fitness center, and business center.

#24

Civil Air Patrol

The **Civil Air Patrol (CAP)** got its start more than 60 years ago flying patrols over America's coastlines, searching for German submarines. In those perilous days it flew 24 million miles, found 173 submarines, attacked 57, hit 10, and sank 2. By Presidential Executive Order, CAP became an auxiliary of the Army Air Forces in 1943 and the USAF in 1947. Today its homeland security mission continues. CAP provided security for the Winter Olympics which was held soon after the September 11 attacks on our nation. It has been asked by NASA to provide surveillance flights above its spacecraft launch sites. CAP performs 95 percent of our nation's inland search and rescue, saving over 100 lives per year. Did you know that it also transports time-sensitive medical materials and flies counterdrug missions?

The **Civil Air Patrol** consists of 64,000 volunteer members flying the world's largest fleet (550) of single-engine, piston aircraft. Divided into 8 geographic regions, 52 wings, and 1700 units, it is all supported by a staff of only 160 based at CAP National Headquarters at Maxwell Air Force Base, Alabama. The CAP also operates the nation's most extensive communications network and has over 1000 emergency services vehicles.

If you're a pilot, or you'd like to be, CAP has plenty of opportunities tailor-made for you. CAP pilots are able to fly its planes for training and performing CAP missions in service to their local communities. If you are 12 to 18 years of age, consider becoming a CAP Cadet. You'll learn plenty about aviation and aircraft and will be given the opportunity to participate in glider and powered-flight training programs.

The best thing about joining a CAP unit is that they're everywhere. There are 1700 of them at strategic locations throughout the United States. If you'd like to become a CAP pilot, you must be at least 17 years of age and hold a valid FAA private, commercial, or airline transport pilot certificate. You must have a Class III or higher medical certificate and a current flight review. You'll have to satisfactorily complete a flight check and pass a written exam. You'll also need to complete aircraft questionnaires for each plane you're qualified to fly. CAP pilots perform some of the organization's most important work.

If you're not a pilot, join now and let CAP help you earn your wings and then become a CAP pilot. This is a great place to meet and work with people who share your interest in flying and want to use their skills in a meaningful way.

Hours

Weekly meetings plus project involvement time

Admission Cost

Dues of $30 per year to region and wing dues

Contact Information

Civil Air Patrol NHQ/DPH
105 South Hansell Street
Maxwell AFB, AL 36112
http://www.cap.gov/
(800) FLY-2338

G/A Airport Serving the Event

Many across the United States

Best Aviation Attraction

#25

Cleveland National Air Show

(Courtesy Cleveland National Air Show.)

The **2004 Cleveland National Air Show** celebrates its 40th anniversary at Cleveland's Burke Lakefront Airport. This coincides nicely with the 75th anniversary of the **Cleveland National Air Races** which used to be held here. Cleveland is the finest regional airshow in the United States. It is well worth attending if you live anywhere within a day's drive.

The top flight demonstration teams in America have and will continue to come to Cleveland. The top military act alternates each year between the Blue Angels and the Thunderbirds. Both are thrilling to watch. The Golden Knights "drop in" at least once every 2 years and often every year.

The active military will let their presence be felt by dispatching line aircraft to this event. Some, like the B-2, will do only a flyby then be gone. Many more will do a flyby, land, and join the other aircraft in the static display area. It is a fine way to see ships like the A-10 Warthog up close. Their agility makes them appear much smaller than they actually are when we see them hitting targets in Afghanistan or Iraq. On television I get the idea that it is no bigger than my Piper Cherokee. When I stand next to it, I am left with only one thought: this is a mighty big aircraft.

Flight museums from as far away as California and Texas will dispatch their period warbirds to Cleveland. You'll see, hear, and touch the aircraft that won our freedom in World War II. Mustangs, Corsairs, Avengers, Flying Fortresses, and Liberators will all be here.

Since August 1999, Burke has been home to the **International Women's Air & Space Musuem.** Exhibits can be seen within the concourse areas. Each year IWASM hosts an annual luncheon the Friday before the air show weekend.

Burke Lakefront Airport would be worth a visit without the pomp of an airshow. Frequently referred to as one of Cleveland's best-kept secrets, Burke's proximity to downtown Cleveland gives the city a strong selling point in its continuing effort to attract new business and to encourage existing businesses to expand locally. The convenience to downtown, ready access to interstate highways, and efficiency of operations make Burke the corporate aviation gateway to Greater Cleveland. Last year, it handled more flight operations than 25 percent of the volume handled at Hopkins. Now that Chicago has destroyed Meigs Field, Burke is the only close-in corporate airport in a major Midwestern city. When corporations or small businesses search for new locations, airport facilities rank among the top three most important considerations. What was Mayor Daley thinking about? Burke's continuance and Meig's demise is one very good reason to locate a business in Cleveland and not Chicago. Both cities have gigantic traffic problems. When it's time to leave a downtown office and jump on a corporate jet, the 7-minute drive to Burke makes much more sense than the afternoon you'll spend trying to get to Chicago's Palwaukee.

Don't just go to this airshow. Be part of it. If you have never been part of the volunteer staff for a major airshow, you have missed out on something exciting and unique. By volunteering you'll have a chance to learn about the entire behind-the-scene action that a mere attendee never sees or even hears about. You'll be there when the major acts arrive. You'll hear about the last-minute problems that crop and you'll be part of solving them. If you would like to volunteer, contact the show's office at (216) 781-0747. Do this quickly as only 200 are chosen for these key positions.

Dates

Annually on Labor Day Weekend

Hours

Gates open at 9:00 a.m.

Flying is from 11:00 a.m. to 4:30 p.m.

Admission Cost

General Admission

Adults	$14 in advance; $16 at the gate
Children (6–11)	$12 in advance; $16 at the gate
Children 5 and younger	Free

Reserved Box Seats

$18 per person regardless of age (Box seats are folding chairs set up along the flight line)

Contact Information

Cleveland National Air Show

1501 North Marginal Road, Suite 166

Cleveland, OH 44114

(216) 781-0747

info@clevelandairshow.com

http://www.clevelandairshow.com/

G/A Airport Serving the Event

Cleveland, Ohio (Hopkins International Airport—(CLE)
www.clevelandairport.com

Cleveland, Ohio (Cuyahoga County Airport—CGF)
(216) 267-3700

FBO Name/Phone Number

Air Services of Cleveland (CLE)
(216) 267-3711

Corporate Wings (CGF)
(800) 261-1115

Ground Transportation

Avis
(216) 265-3702

Best Place to Stay $$$$$

Glidden House Inn
1901 Ford Drive
Cleveland, OH 44106
(216) 231-8900
www.gliddenhouse.com

The Glidden House features 52 beautiful and spacious rooms along with 8 suites in the original mansion. Both possess a delicate blend of originality and quality that will appeal to the most discerning of tastes.

Best Place to Stay $

Days Inn Cleveland Lakewood
12019 Lake Avenue
Lakewood, OH 44107
(866) 223-9330
www.daysinnhotelreservations.com

Best Aviation Attraction

#26

Cradle of Aviation Museum

The Cradle of Aviation Museum is unique for two reasons. It is America's newest major aviation museum and I believe its most spectacular one. I much prefer it to the Smithsonian's **National Air and Space Museum.** It sets out to tell the tale of aviation in Long Island and in doing so chronicles the history of flight, for the story of one belongs to the other. It is the only museum in the United States other than the **National Air and Space Museum** where the entire story of flying is told from the early 1900s until tomorrow.

To my mind there are three major aviation achievements and only one was not directly connected to this geography. If it was the Wright Brothers who taught man how to fly, it was most certainly Lindbergh who showed us why we should. On May 20, 1927, he took off from a spot near the museum, Roosevelt Field. The actual ground he struggled to break free of is today the asphalt covered parking lot of a shopping center. Ask one of the museum's docents for directions. You need to stand where he stood and imagine his takeoff. Lindbergh took the American imagination on a trip from Long Island to Paris. His flight taught us that aviation was about high-speed travel. The world, all of it, now belonged to each of us.

If Lindbergh gave us a reason to fly, then it was Neil Armstrong who removed the boundary. Landing a ship manufactured near here was indeed "one small step for man and one giant leap for mankind." With it, he gave us the stars. Yes, Houston, the Eagle had landed at Tranquility Base and in a very real way it had taken off from *Long Island—The Cradle of Aviation.*

In the span of the past 100 years aviation was called upon to defend our very freedom to dream that we might go to the stars. Many of the airplanes that went to war were built here also. Two major companies answered the call and became legends to thousands of thankful flyers. Grumman became simply "the ironworks" to Naval aviators who flew their faithful planes. You know them; they were the ones that brought our men back aboard ships with their wings shot so full of holes that they should have crashed into the ocean. But they flew on because of the way the men of Long Island designed and built them—like iron angels!

If you can come to only one museum make it this one. It isn't just about the planes and spacecraft that are displayed; it is about the place that made them. When you come here, you will be overpowered by the sense of history that surrounds you. History is a combination of time, people, things, and place. Here they all come alive.

Come with a sense of curiosity and leave with a sense of wonder.

Dates

Every day

Hours

Wednesday through Sunday	10 a.m. to 5 p.m.
Monday and Tuesday	10 a.m. to 2 p.m.

Due to limited capacity, timed tickets may be issued.

Admission Cost

Adult	$7.00
Child (2–14)	$6.00

Contact Information

Cradle of Aviation Museum
Charles Lindbergh Blvd.
Garden City, NY 11530
(516) 572-4111
(516) 572-4079 (fax)
http://www.cradleofaviation.org

G/A Airport Serving the Event

Farmingdale, NY (Republic Field—FRG)
Phone: (631) 752-7707
Tower: 188.8
ASOS: (631) 752-8129
Runways:

1–19	5516′ × 150′
14–32	6827′ × 150′

FBO Name/Phone Number

Atlantic Long Island
(631) 752-9022
(631) 752-9099 (fax)
www.atlanticaviation.com

Ground Transportation

Rental car

Best Place to Stay $$$

Holiday Inn
215 Sunnyside Blvd.
Plainview, NY 11803
(516) 349-7400

Best Place to Stay $$$

Hampton Inn Long Island/Commack
680 Commack Road
Commack, NY 11725
(631) 462-5700

Best Aviation Attraction

#27

Glenn H. Curtiss Museum

(Courtesy Glenn H. Curtiss Museum.)

Glenn Curtiss is a complicated figure. I like him for the work he did, particularly for the wonderful aircraft he produced and the company he built. My all-time favorite airplane is the P-40 Warhawk which emerged from his firm's design table. He personally produced more than 500 inventions in his lifetime. The distraction for me was his obsession with winning at any cost and his early uncompensated infringement on the Wright Brothers patents.

Aviation would not have gotten off the ground as quickly as it did without Curtiss. His early flying days were monumental. It was Curtiss who designed and flew the first flying boats. The world took great advantage of these designs. During World War I it was Glenn H. Curtiss who gave America the "Jenny" and it was the Jenny and the "barnstormers" who flew her, that *really* introduced America to the airplane!

This museum profiles this American life and brings together a fine collection of the things that his life touched and produced. It is worth the trip, for without Curtiss the airplane would have developed much more slowly. Next, it is worth the trip for the geography it occupies. Can there be a more beautiful place on earth than the Finger Lakes region of New York State? I think not. Your stay at the museum may be brief, but you'll want to spend at least a week in the area.

Dates

Open	Daily
Closed	Easter Sunday, Thanksgiving Day, Day before Christmas, Christmas Day, and New Year's Day Monday through Wednesday in January, February, and March

Hours

Monday through Saturday (May 1–October 31)	9:00 a.m. to 5:00 p.m.
Monday through Saturday (November 1–April 30)	10:00 a.m. to 4:00 p.m.
Sundays (May 1–October 31)	11:00 a.m. to 4:00 p.m.
Sundays (November 1–April 30)	12:00 a.m. to 5:00 p.m.

Admission Cost

Adults	$6.00
Seniors (65 and over)	$4.50
Students (7–18)	$3.50
Children (6 & under)	Free
Members	Free

Contact Information

Glenn H. Curtiss Museum
8419 Route 54
Hammondsport, NY 14840
(607) 569-2160
(607) 569-2040 (fax)
http://www.linkny.com/~~curtiss/
trafford@linkny.com
curtiss@linkny.com

G/A Airport Serving the Event

Penn Yan, NY (Penn Yan Airport—PEO)
Phone: (315) 536-4471
CTAF-Unicom: 123.0
AWOS: (315) 536 4102
Runways:

1–19	3265′ × 50′
10–28	4500′ × 100′

Using this airport may just make the trip that much better for you. This is the home of **Penn Yan Aero,** builders and distributors of some of the finest aircraft engines flying. If it's time to repower your ship, this is a fine place to do it. For details have a look at their Website: www.pennyanaero.com.

FBO Name/Phone Number

Seneca Flight Operations
(607) 536-4471

Ground Transportation

Rental car

Best Place to Stay $$

Village Tavern Restaurant & Inn

P.O. Box 573
Hammondsport, NY 14840
(607) 569-2528
(607) 569-3560 (fax)

There aren't *any* motels or traditional hotels in Hammondsport, *none!* There are a few bed and breakfast places and this is one wonderful European-style inn. It is much like something you might find in England: rooms upstairs and a restaurant and pub downstairs. It is fun and it is located on the Hammondsport Village Square. Don't just visit Hammondsport; live there for a few days. That's the difference between a chain motel and a tavern on the Square! You'll thank me for this one!

Best Aviation Attraction

#28

Dayton Air Show

(Courtesy The AeroShell Aerobatic Team.)

Dayton, Ohio, is the hometown of Wilbur and Orville Wright, the inventors of powered flight, and has long been considered the birthplace of aviation. The city and the surrounding region offer many aviation-related attractions and historical sites as well as a thriving downtown, scenic parks, and high-quality shopping and dining. When you come out for the **Dayton Airshow,** plan to spend a few extra days to explore the area and to get up close and personal with the land that gave us wings.

The **Dayton Air Show,** the world's largest air show, is produced by The United States Air and Trade Show. In 2003 Dayton drew especially large crowds for the flight centennial celebration in the hometown of the Wright brothers. No other air show has ever had *all* three top-flight performance teams. This year the fans were able to witness the **U.S. Air Force Thunderbirds, U.S. Navy Blue Angels,** and **Canadian Forces Snowbirds** performing on the same day in the same sky. In addition, more than 150 unique planes were on display at the 130-acre show site that is the Dayton International Airport.

Dates

July 17 and 18, 2004

Hours

Gates open	8:00 a.m.
Flying begins	9:00 a.m.
Flying stops	6:00 p.m.

Admission Cost

General Admission: Adults	$19
General Admission: Youths (ages 6 to 11)	$16
General Admission: Senior citizens	$16
General Admission: Ages 5 or under	Free
Parking	$7

Website

http://www.usats.org

Contact Information

United States Air and Trade Show
P.O. Box 460
Dayton International Airport
Vandalia, Ohio 45377-0460
info@airshowdayton.com
(937) 898-5901
(937) 898-5121 (fax)

G/A Airports Serving the Event

There are three principal airports to consider if you are planning on flying into the Dayton Air Show. Naturally, **Dayton International** is the first choice. This is the site of the air show. It is also a tremendous airport with multiple IFR landing procedures including ILS. Be warned that parking is limited. Advanced planning is the key. Call the airport office and make a reservation. They'll want to know your arrival date, the approximate time, and the duration of your stay. Do this as early as you possibly can!

Second choice is the **Wright Brothers Airport,** which is located in Miamisburg, Ohio. It is almost 20 miles away and you will have to rent a car. The longest runway is a comfortable 5000 footer. There is only one published IFR approach, a VOR/DME so you can get in should the weather turns soupy but you'll need higher minimums than at Dayton International. Call the airport, reserve a parking spot and a rental car. You'll need both.

The best of the remaining choices is **Greene County Airport** in nearby Xenia. There are no published instrument approaches. The single runway is almost 4000 feet long. Again, call to reserve ramp space and a rental car.

There are many other airports in the area. The three mentioned here are merely my opinions of the best for the average aviator flying a four-place single or bigger with tricycle landing gear.

Dayton International
Phone: (937) 454-8200
Runways: 10,901′ × 150′, 8500′ × 150′, 700′ × 150′
Identifier: DAY

FBOs

Stevens Aviation, Inc.
(937) 454-3400
Wright Brothers Aero
(937) 890-8900

Ground Transportation

Enterprise Rent A Car
(800) Rent A Car

Wright Brothers Airport
Phone: (937) 885-3662
Runway: 5000′ × 100′
Identifier: MGY

FBOs

Aviation Sales
(937) 885-3662
Commander-Aero, Inc.
(937) 885-5580

Ground Transportation

Enterprise Rent A Car
(800) Rent A Car

Greene County Airport
Phone: (937) 374-2650
Runway: 3947′ × 75′
Identifier: I19

FBO

Greene County Airport Authority
(937) 376-8107

Ground Transportation

Enterprise Rent A Car
(800) Rent A Car

Best Place to Stay $$$

Doubletree Dayton Downtown
11 South Ludlow
Dayton, OH 45402
(937) 461-4700
(937) 461-3440 (fax)
http://www.doubletree.com

The Doubletree Dayton Downtown is the only historical luxury hotel in Dayton. The concierge level offers enhanced amenities, robes, and a private lounge serving complimentary continental breakfast and evening hors d'oeuvres and beverages in a club room atmosphere. Plan on making a visit to the **Wright Brothers Bicycle Shop,** a short walk away, or better yet, rent a bike!

Best Place to Stay $$

Best Western Executive Hotel
2401 Needmore Road
Dayton, Ohio 45414
(937) 278-5711
(937) 278-6048 (fax)

Best Aviation Attraction

#29

Drop Zone

The most exciting experience you can imagine is jumping from a perfectly good airplane at an altitude of 13,500 feet or so, free-falling and then gliding gently back to earth under a high-tech multicolored canopy that flies like a wing. This is skydiving.

Get started by selecting a fully accredited school. The National Skydiving Association provides a list on the Internet of all of the schools they have accredited in the United States. It would be a good idea to visit the school, meet the instructors, and watch a student or two make the plunge before you do. Make sure this is the place you want to trust with your life because that's what you're about to do.

If you can't find a suitable school near you, I would recommend strongly that you contact **Skydive DeLand.** It has long been recognized as a worldwide leader in student training, competition training, and equipment development. DeLand, Florida, is a short distance north of Orlando, Florida. A trip to the Skydive center can easily be combined with a family outing to Disneyworld, or Universal's theme parks.

To begin skydiving here, you may start with a tandem jump or go right into the free-fall training program, known as *accelerated freefall (AFF)*. These training methods were developed at DeLand and are practiced throughout the world. Most people begin with the tandem jump. You and your instructor will share a tandem harness, free-falling from 13,500 feet at speeds exceeding 100 mph. After free-falling, your instructor will open the dual-sized parachute, and together you will fly back to the ground for a soft landing. You'll be higher than a kite and your wallet will be about $160 lighter. What should you wear for the jump? Tennis shoes and comfortable clothing. The school will provide you with a jump suit.

You'll get to 13,500 feet aboard a fast-climbing jump plane with a group of other jumpers, pros, and first timers. Most schools use twin engine aircraft like the DeHavilland Super Twin Otter. Right away you'll notice that it isn't going to be a luxury cruise as there are no seats in this bird. Everybody sits on the floor. The door is probably not installed, so there is just this huge hole in the fuselage. As you climb, the wind whips through it and you'll swear that the hole just gets bigger and bigger the higher you climb. Second thoughts? Everybody has 'em. Jumping out of this airplane is going to be a difficult moment. For some people it isn't at all, they just know that this is not for them.

If you're one, skip the tandem jump process and go AFF. It is a seven-jump course that will make you a very competent skydiver at the end. In this program the first course is preceded by a little more than half a day of ground school. Then you'll make your first jump—solo. Well, not completely solo, two instructors will jump with and assist you on the way down. It is a good feeling to know they're nearby. The feeling of flying all by yourself is amazing. The entire course will run you about $1000 with all equipment supplied by the school.

Me? No, I have never jumped and imagine that I never will. Watching these folks return to earth gently supported by skill and nylon is a beautiful sight. I do admit to being a frequent and enthusiastic spectator.

Dates

Daily

Hours

Sunup to sundown

Admission Cost

$175.00

Contact Information

National Skydiving Association
http://www.skydivinginformation.com/
Skydive DeLand
1600 Flightline Blvd
DeLand, FL 32724
(386) 738 3539
info@skydivedeland.com
http://skydivedeland.com

Best Aviation Attraction

#30

Edwards Air Force Base Open House

The **Air Force Flight Test Center** hosts the **Edwards Air Force Base Open House and Air Show** annually in late October. More than 400,000 people attend and enjoy a weekend showcasing the capabilities of the Air Force, AFFTC through assigned and visiting aircraft flying demonstrations, static displays, and attractions.

"We have the world's most unique aircraft as part of our normal operations. Few air shows have the variety of aircraft we have at Edwards," said Maj. Kevin Steffenson, the air boss of the 2003 show. He's right and that's why nearly half a million people show up in 2 days. Many of the planes you'll see here are not shown anyplace else, ever!

This airshow is unique in that it is held at one of aviation's most hallowed sites. This is where the sound barrier was cracked wide-open. The X-15 left from here many times to fly to the edge of space and to return. When the weather is bad in Florida, this is where the space shuttle returns to earth. But this is not just a ghosttown from our past; advances are being made here everyday. This is still the Air Force Flight Test Center. Not only are new designs run through the ringer here, top pilots are as well. When they finish they are test pilots, for this is the home of the Air Force's test flight school. If there was no air show it would be worth the trip, just to stand on this ground, look at these hangars and planes, and smell the jet fuel-laden air. This is where you *must* come!

The air show is usually highlighted by the Thunderbirds, the Air Force's flight demonstration team. Many military and civilian aircraft will be flown and displayed. In 2003 the newest in our air arsenal were displayed in a hangar: F-18, F/A-22 Raptor, the X-35 Joint Strike Fighter, and the pilotless Global Hawk and Predator. The flightline was filled with civilian and military planes, including the B-1B Lancer, B-25 (1944), B-52H Stratofortress, CASA Jet (HA 200), C-12, C-17A Globemaster III, C-46, CH-46E Sea Knight, CH-47 Chinhook, CH-53E Sea Stallion, EW-9D, F-5 Tigershark, F-15E Strike Eagle, F-15C Eagle, F-16 VISTA, F-16 Falcon, F-86, KC-135R (or KC-135 Stratotanker), KISS FM Helicopter (civilian aircraft), LAPD A-STAR Helicopter (civilian aircraft), L-5 Stinson, L-9 Stinson, L-19 Birddog, L-29, L-39 Albatross, MIG 15, MK-16, NASA 747, NATO AWACS, PT-17, SERO, TS-A, TS-11 ISKRA (Polish), T-6A II, T-28C, T-33, T-34A Mentor, UH-60 Blackhawk, YAK 52 (Russian), VISTA, YAK-3 UA (Russian), AT-6 Texan, B-2 Spirit, F-86 Saber Jet, F-117A Nighthawk, Glider, N-9M, P-51, Thunderbirds (F-16), TOW, U-2, YA-10B.

Dates

Late October annually

Hours

Gates open at 7 a.m.
Flying begins at 9:30 a.m. both days.
Show closes at 4:30 p.m. both days.

Admission Cost

Free
No parking fees

Website

http://www.edwards.af.mil/oh.html

Contact Information

Air Force Flight Test Center
Public Affairs
AFFTC/PA
1 S. Rosamond Blvd.
Edwards AFB, CA 93524
Public Affairs: (661) 277-3510

G/A Airport Serving the Event

Mojave, CA (Mojave Airport—MHV)
The Nation's Civilian Flight Test Center
Website: http://www.mojaveairport.com/
E-mail: stuart@mojaveairport.com
Phone: (661) 824-2433
CTAF-Unicom: 127.6
Runways:

12–30	9500′ × 200′
8–26	8000′ × 100′
2–22	4700′ × 50′

Arresting gear is installed on both ends of runway 12–30! Transient aircraft tiedowns are available and highly recommended due to the robust winds adjacent to the airport administration building. There is no tiedown fee.

FBO Name/Phone Number:

East Kern Airport District
(661) 824-2433

Ground Transportation

Enterprise Rent A Car
800 Rent A Car

Best Place to Stay $$$$$

Mariah Country Inn and Suites
1385 Highway 58
Mojave, CA 93501
(866) 627-4241
http://www.mariahhotel.com/
reservations@mariahhotel.com

This is a AAA three-diamond-rated motel. It is the nicest facility for many, many miles. Naturally, they have a pool, spa, and workout room. They operate a shuttle to and from the airport. The Mariah Inn restaurant, which is attached, is one of the best in town.

Best Place to Stay $

Econo Lodge
2145 SR 58

Mojave, CA
(661) 824-2463
http://www.econolodge.com/

This is a AAA two-diamond-rated facility. It is clean, comfortable, and affordable. There is a pool and pets are accepted. The bonus feature is that each room is equipped with a microwave and refrigerator.

#31

Eglin AFB Open House & Air Show

The **Eglin AFB Open House & Air Show** in Florida is one of the truly great regional airshows. Naturally The **Thunderbirds** will headline the show. It is worth the trip just to watch them fly. The usual assortment of other acts and static display will be in attendance. The **Golden Knights** will probably drop in as well.

Like so many of the airshows in Florida one of the things that make them so worthwhile and so well attended is their location. Particularly the ones that happen in the spring, long before the Florida sunshine turns the state into a sauna. Eglin AFB has the distinction of being located in the middle of 100 miles of gorgeous powder white sand-covered beaches. The waters of the Gulf of Mexico are absolutely gin clear here. Its proximity to Alabama and Georgia has earned it the nickname "Red Neck Rivera." I hope the name sticks and keeps the more sophisticated types from discovering this near paradise.

My advice for a weekend military base air show is double emphasized for this one. Don't go to the show. Instead show up for the preshow practice session on Friday. It kicks off midmorning and runs through late afternoon. The Thunderbirds will practice in the morning and the afternoon. All other acts will practice at least once.

Follow this strategy and go to the beach on Saturday and Sunday. The only thing you'll miss is the crowd. Eglin AFB is so close to the ocean that you'll see a lot of the weekend action anyway.

Dates

April 3–4, 2004 (early April in even-number years)

Hours

9:00 a.m.–4:30 p.m.

Admission Cost

Free

Contact Information

Eglin AFB Open House & Air Show
101 West D Avenue, Suite 110
Eglin AFB, FL 32542-5498
(850) 882-3931 ext. 484
http://www.eglin.af.mil/Eglin_Airshow/

G/A Airport Serving the Event

Destin, FL (Destin-Fort Walton Beach—DTS)
Phone: (850) 837-6135
CTAF-Unicom: 122.8
AWOS: (850) 654-7128
Runway: 14–32 5000′ × 100′

FBO Name/Phone Number

Miracle Strip Aviation
(850) 837-6135

Ground Transportation

Rental car

Best Place to Stay $$$

Holiday Inn—Destin
1020 Hwy. 98 E.
Destin, FL 32541
(850) 837-6181
hidestin@gnt.net

Best Aviation Attraction

#32

Evergreen Aviation Museum

(Courtesy Evergreen Aviation Museum.)

The centerpiece or "crown jewel" of the **Evergreen Aviation Museum** is a plane known around the world as the *Spruce Goose.* This is the plane that caused such a huge problem for Howard Hughes with the U.S. government. Was it this aircraft and those problems that finally drove the reclusive billionaire into a lifestyle that only Saddam Hussein could appreciate? I don't know the answer to that question; neither does anyone else. I can tell you that this aircraft is worth seeing. It is unique from many perspectives.

Other than the *Spruce Goose,* what makes the **Evergreen Aviation Museum** special enough to be selected as one of the *101 Best Aviation Attractions?* It is curatorial vision. Many, some would say too many, museums concentrate on one historical period or one segment of aviation. The obvious example is World War II military aircraft, *Warbirds* as they are affectionately known. So many museums have been built around them that they have become trite and boring. This museum is refreshing in its approach.

At least one aircraft from every historical period and every aviation segment is on display. It isn't always possible to get an actual survivor so replicas are used when necessary. The Wright Flyer is an example. No museum has the original Wright Flyer. It crashed and was damaged beyond repair on its fourth flight of December 17, 1903. The one presented here is a replica and so is the one in the austere Smithsonian. You can get close enough to this one to appreciate the genius of the Wright design. Making a 750-pound machine powered by a 12-horsepower home-brewed engine fly is a pretty fair trick. Every airplane since has relied on the same control principles.

My favorite military airplanes are here. Each is from a very different period. There is a Curtiss *Jenny* from World War I. The P-40 *Warhawk* of World War II was made famous by the Flying Tigers. It is a delight to stand next to one. The German aces of the same era preferred the ME-109. The one you'll see is listed as flyable! From the modern era you'll see an F-15 guarding the front gate.

Private civilian planes are in abundance. The ones I care about are here. Everyone's favorite is the J-3 Cub sporting its telltale yellow and black paint job. A good choice to be sure, but it is the 1947 Beechcraft Bonanza 35 that proves the excellence of this curation. It is the plane that modern general aviation is built on. Every newly minted private pilot dreams of someday owning a Bonanza.

The plane that made commercial travel by aircraft a reality is here, a near-perfect 1936 DC-3. Close by is an early attempt at opening ticket counters across the world, a 1928 Ford Trimotor. Look closely. Is it really possible that these two designs are separated by only 8 years?

Even homebuilt experimental craft are displayed here. They should be because 25 percent of annual aircraft production comes from this sector. The choices are perfect. You'll see a Bede BD-5B and a Baby Great Lakes.

When you leave this museum you will have been touched by aviation's full spectrum. Best of all, your visit is capped by the *Spruce Goose.* Stand under its giant wing and look up. At 320 feet, it is the longest wing ever built! They support eight supercharged 28-cylinder engines generating 3000 horsepower. The eight of them equal the power of seven locomotives of their day; they comprise the most piston-horsepower ever placed on an airplane.

Dates

Open daily
Closed Thanksgiving, Christmas and New Year

Hours

Winter Hours	9:00 a.m.–5:00 p.m.
Summer Hours	9:00 a.m.–6:00 p.m.

Admission Cost

Adults	$11.00
Seniors (65+)	$10.00
College students (with ID)	$7.00
Children/Youths (6–18 years)	$7.00
Members	Free
Children (0–5 years)	Free

Contact Information

Evergreen Aviation Museum
3685 NE Three Mile Lane
McMinnville, OR 97128

(503) 434-4180
(503) 434-4058 (fax)
http://www.sprucegoose.org

G/A Airport Serving the Event

McMinnville, OR (McMinnville Municipal—MMV)
Phone: (503) 434-7411
CTAF-Unicom: 123.0
ASOS: (503) 434-9153
Runways:

17–35	4676′ × 150′
4–22	5420′ × 150′

For pilots, this is a great museum to visit. You can land at McMinnville and park at Cirrus Aviation. The museum is just a short walk across the street. You really won't need a car or a motel unless you arrive late in the day and plan to visit the museum the next day.

FBO Name/Phone Number

Cirrus Aviation
(503) 472-0558

Ground Transportation

Rental car

Best Place to Stay $$

Red Lion Inn & Suites McMinnville
2535 NE Cumulus Avenue
McMinnville, OR 97128
(503) 472-1500
(503) 474-1171 (fax)

Best Aviation Attraction

#33

Fantasy Flight Camps

Start with terrific facilities, add a seasoned, well-versed subject expert, stir in a few eager participants, and top it off with a ride on the subject of the weekend course—a perfectly restored antique aircraft. Do it four times a year. What do you have? The EAA's deservedly popular **Fantasy Flight Camp** program! Save your money, collect Coke cans by the side of the road, do whatever you have to do to get the cash for this experience. It is terrific. Every effort is made to make this a true "fantasy" experience for the aviation enthusiast, including privileged access to EAA facilities and specialists. You're an honest to goodness VIP this weekend!

Let's say you're really interested in Lindbergh and the *Spirit of St. Louis.* Imagine spending a weekend as a VIP in Oshkosh, WI, at EEA's Wittman Field Headquarters. An EAA staffer will meet your flight and whisk you away to your weekend digs. Saturday morning you'll meet with your class at the EAA Aviation Museum. A special instructor has been selected for his thorough knowledge of this subject. With slides, films, maps, and a slew of museum artifacts the instructor will fill you in, until you feel that you knew Lucky Lindbergh and personally escorted him across the pond. All around you are other folks who have a burning interest in the *Spirit of St. Louis* just as you do. You'll doubtless make a friend or two. Flying buddies after all are hard to come by and are treasured finds.

Just when you thought it couldn't get any better, you'll be escorted to the ramp to look at, touch, and sit inside a replica of *The Spirit of St. Louis.* This one is different than all the others for two reasons. The first is that it is perfect. The more important is that it has two seats, one for you and one for your instructor pilot. By the way, your instructor has over 1000 hours in type. You're going for the ride of a lifetime! It will be at once the longest and shortest 30 minutes that you've ever experienced. Bring a good camera and a good camcorder.

Now if you happen to be a pilot you'll be pleased to know that the time you spend in this baby with your instructor is loggable as PIC. Better yet somehow the EAA restoration team wound up with the exact tail number of Lindy's ship, N-X-211. So that's the designation that will go in your logbook!

As soon as it began, your **Fantasy Flight Camp** is over. But that's O.K. because the EAA does this four times a year and each camp is different! Don't be a pig though, spread these adventures out over a few years to really feel the effect.

The program fee includes all instruction, materials, meals, and lodging in the **EAA Air Academy Lodge,** double occupancy with shared bathroom facilities, and your flight experience. A limited number of rooms with private baths are available for an additional fee. A nonlodging option is also available. The course is always scheduled over a weekend. All on-site transportation is provided in an EAA vehicle.

Contact EAA for the current program and schedule. In 2003, the camps and fees were as follows:

Spirit of St. Louis Fantasy Flight Camp

$450 EAA Member

$495 Non-EAA Member

Aviation Photography Fantasy Flight Camp

$450 EAA Member

$495 Non-EAA Member

Ford Tri-Motor Fantasy Flight Camp

$450 EAA Member

$495 Non-EAA Member

B-17 Fantasy Flight Camp

$700 EAA Member

$745 Non-EAA Member

Contact Information

EAA AirVenture Museum

PO Box 3065

Oshkosh, WI 54903-3065

(800) 236-4800 Ext. 6820

airacademy@eaa.org.

http://www.airventuremuseum.org/flightops/fantasycamp/

G/A Airports Serving the Event

Oshkosh, WI (Wittman Regional Airport—OSH)

Wittman is a truly wonderful general aviation airport. It offers four runways, the longest of which is 8002 feet. Every IFR approach including an ILS is available.

FBO Name/Phone Number

Orion Flight Services, Inc. (OSH)

(866) 359-6746

Ground Transportation

City	Rental Car Company	Phone
Oshkosh	Hertz Rent-A-Car	(800)654-3131
	Avis Rent-A-Car	(920)730-7575
Appleton	Hertz Rent-A-Car	(800)654-3131
	National Car Rental	(920)739-6421
	Enterprise Rent-A-Car	(800)RENT-A-CAR

Best Place to Stay Oshkosh $$$

Hilton Garden Inn Oshkosh

1355 West 20th Avenue

Oshkosh, WI 54902

(920) 966-1300

(920) 966-1305 (fax)

This premier hotel is right on the Wittman Regional Airport grounds next to the **EAA AirVenture Museum.** Plan on spending 2 days here, because you'll need at least that much time!

Best Place to Stay Appleton $$$

Hilton Garden Inn Appleton

720 Eisenhower Drive

Kimberly, WI 54136

(920) 730-1900

(920) 734-7565 (fax)

Best Aviation Attraction

#34

Fantasy of Flight

Fantasy of Flight in Florida is the world's largest private collection of vintage aircraft. The displayed aircraft include a Curtiss Jenny, a Ford Tri-Motor, German JU-52, Japanese Zero, Corsair F4U, B-17 Flying Fortress, and B-24 Liberator. The centerpiece of **Fantasy of Flight's** fleet is a huge Short Sunderland, the last airworthy four-engine civilian flying boat in the world! The facility encompasses much more than the aircraft. The Museum reflects the airports of aviation's Golden Era, the late 1930s and early 1940s. The familiar red-and-white checkerboard paint of the water tower and the period menu of the operational **Compass Rose Diner** recall a small airport of more than a half-century ago.

If this were just another display of vintage aircraft it would probably not be worth seeing. What makes **Fantasy of Flight** one of the *101 Best Aviation Attractions* is the thoughtful effort the staff uses to involve each visitor in aviation and to bring the aircraft of their collection to life. Each day at 2:30 p.m. the Aircraft of the Day Demonstration begins. One of the pilots will fly a vintage aircraft from the display hangar. The pilot will give a brief lecture on the aircraft and its history. Following a question-and-answer session the pilot will take the aircraft to the skies over **Fantasy of Flight** and make history come to life!

At 12:45 p.m. the Restoration Shop Tour starts in the restoration shop. Skilled craftsmen will explain and demonstrate the techniques they use, including metal shaping and smoothing, and frame building. In the warehouse there are outstanding aircraft specimens including a Navy Panther Jet and a German JU-52. Engines dating back to the early 1900s, including a collection of never-used Allison engines, are stored here also. The south hangar houses the machine shop where vintage engines and various metal parts no longer produced or available are produced. Look in on the wood shop where wooden wings are covered with fabric and resin in what is called *the doping process.*

It seems that everyone's favorite World War II fighter is the P51. There is a fine example of one here. Every hour a video presentation of the P51C is given.

Visitor involvement is the watchword of Fantasy of Flight. After you've seen and listened and watched it's time to do! Start with the full-motion Fightertown flight simulators where you can virtually participate in a battle over the South Pacific! Open 9:30 a.m. to 5:00 p.m. Unlimited use is included in your admission! Conclude your visit with a ride in a period biplane. This experience is only available during the winter season, November 1 through April 30, and there is an additional charge of $54.95 per person for this 18-minute flight.

Dates

Open everyday
Closed Thanksgiving and Christmas

Hours

9 a.m.–5 p.m.

Admission Cost

Adults	$24.95
Seniors (Ages 60+)	$22.95
Children (Ages 5–12)	$13.95
Children (under 4)	Free

Save $3.00 on any admission by downloading a coupon at the Fantasy of Flight Website (http://www.fantasyofflight.com/coupon.htm).

Contact Information

Fantasy of Flight
1400 Broadway Blvd. S.E.
Polk City, FL 33868
(863) 984-3500
http://www.fantasyofflight.com

#35

Farnborough International Exhibition and Flying Display

The **Farnborough International Exhibition and Flying Display** is a biennial event staged at the Farnborough Aerodrome in the United Kingdom in even-numbered years. The next exhibition will be held from July 19 to July 25, 2004. Like the **Paris Airshow,** which is staged on approximately the same dates in odd-numbered years, this is really a trade show. It is about the business end of aviation. Airlines and defense departments from around the world come to check out the latest gear. Airframes, avionics, and armaments are bought and sold here.

Farnborough bills itself as the world's premier aerospace business event. It continues to please its exhibitors and visitors. Business is done here, a lot of it. In 2002 the value of orders announced at the show reached $9 billion, well over $100 million a day.

It is the largest *temporary* exhibition in the United Kingdom, if not the world. In 2002, 1260 exhibitors from 32 countries were accommodated in 71,000 square meters of temporary structures and 42,000 square meters of outdoor space. Attendance reached 290,000 visitors over the 5 trade and 2 public days.

In the air, 133 displays were flown by 79 aircraft from 14 countries. Thirty-nine military and 40 civilian aircraft participated. There were 356 helicopter and 763 aircraft movements. The lightest aircraft weighed in at 680 kg with the heaviest at 188,000 kg.

Many people plan their vacations around this airshow. You may want to as well. Europe goes on vacation in August. So many sites and services become unavailable during this, their hottest month. You would be best served by planning your trip on the front end of the show rather than trying to tour after the show. It is a perfect time to be in the United Kingdom and Europe. London offers a wide choice of hotels, easy rail transportation, and an abundance of post-airshow entertainment. Staying outside London is less expensive but there is less to do.

The most efficient way to travel to the show is by rail, departing from London's Waterloo Station and traveling to Farnborough's main station. The journey time is approximately 35 minutes, and there are at least three trains per hour.

Dates

The public may attend on the final weekend of the show.

Hours

Exhibits	9:30 a.m. through 6:00 p.m.
Airshow	5:00 p.m. through 5:00 p.m.

Admission Cost

£20 advance purchase
£25 at the event

Contact Information

Society of British Aerospace Companies
Duxbury House
60 Petty France
Victoria
London SW1H 9EU
United Kingdom
Tel: 44 (0) 20 7227 1043
Fax: 44 (0) 20 7227 1039
farnborough@sbac.co.uk
http://www.farnborough.com/

Best Place to Stay $$$$$

Claridge's
Brook Street
Mayfair
London W1A 2JQ
United Kingdom
(800) 637-2869 (toll free from North America)
44 (0) 20 7950 5481
www.savoy-group.co.uk/claridges/default.asp

Best Place to Stay $

Comfort Inn Victoria
18-24 Belgrave Road
London EN, GB, SWIV 1QF
United Kingdom
(44) 207 233 6636
(44) 207 932-0538

Best Aviation Attraction

#36

Fleet Week San Diego

On the first weekend of October at about midday San Diego Bay and the sky above it host the world's most spectacular parade. **Fleet Week Sea-n-Air Parade** is the kickoff for **Fleet Week San Diego.** The U.S. Navy will parade many types of ships and aircraft that comprise the world's mightiest Navy. An Aircraft Carrier, Cruiser, Guided Missile Destroyer, Destroyer, Frigate, Nuclear Submarine, Amphibious Assault Ship, Amphibious Dock and Platform Landing Ships, SEAL Support Boat, and other small craft are normally all present. The public can view the parade from many spots along the waterfront.

The might of the ships is awesome but the thrill of watching the Blue Angels scream above them at near mast height is stunning. Often they will come over the ridge that separates San Diego from La Jolla unannounced until the sound of their engines shakes the ground beneath your feet. They continue to hug the ridge as they roar toward the Bay and the ships on parade. Thus setup for a perfect low pass, they hold a diamond formation until they approach San Diego's famous Coronado Bay Bridge. In an instant, the four birds of the diamond are joined by Blue Angels five and six as all transition to near vertical flight on their way to 15,000 feet and a perfect performance of their world famous star burst.

The *best* place to see this show isn't the Del Coronado even though I love that hotel. You'll want to be in one of two bay-fronting semicircular buildings that comprise the **San Diego Marriott Hotel & Marina.** The views from their bay-facing rooms are spectacular. Book early and get one of them! Pay the price!

The Blues aren't the only aircraft that you'll see coming down the shoot this day. All are seen perfectly from your hotel room or one of the special viewing areas that the hotel will set up for guests. I have heard that they even open the roof top for viewing. If that's true—*Wow* is the only word that would fit the experience!

Dates

Early October or late September each year

Hours

Early afternoon, between noon and 3:00 p.m.

Admission Cost

Free

Contact Information

http://www.fleetweeksandiego.org/

G/A Airport Serving the Event

San Diego, CA (Gillespie Field—SEE)

Phone: (619) 596-3900

Tower: 120.7

ATIS: (619) 448-1641

Runways:

17–35	4147′ × 100′
9R–27L	2737′ × 60′
9L–27R	5341′ × 100′

San Diego is blessed with three great general aviation airports, Gillespie, Brown, and Montgomery. I picked Gillespie only because it has some of the San Diego Air Museum's facilities on board and you might enjoy a visit. All are great fields that will give you an equal experience. San Diego is aviation, so you can't go wrong at any of its airports.

FBO Name/Phone Number:

Royal Jet

(619) 448-4200

Ground Transportation:

Rental car

Best Place to Stay $$$$$

San Diego Marriott Hotel & Marina

333 West Harbor Drive

San Diego, CA 92101

(619) 234-1500

I love the Del Coronado and its beach, but it is not the place to watch this parade. Top honors truly go to the Marriott. Be warned! It is an expensive property and its bay-view rooms will be in high demand for this event. *Book early!*

When the air part of the show is over, grab a really nice ferry boat from the marina at the hotel to speed you across the bay to Coronado Island. Many of the ships, particularly the aircraft carrier, that were in this parade will moor on the Island. Many will be available for inspection. Typically, the carrier will be one of those. Don't try to use your car to drive across the bridge. It and the island's streets will be jammed!

The Naval Air Station at North Island will have many events during Fleet Week. The public will be welcomed at a few of them. Get the schedule and go. This is a historic field filled with amazing warplanes.

Best Aviation Attraction

#37

Fleet Week San Francisco

Early October will find the Blue Angels in San Francisco for what most consider to be the world's best venue for an airshow. They perform over the bay which wraps around San Francisco with the renowned Golden Gate Bridge as a picturesque backdrop. The Blues take over San Francisco's skies on the Thursday preceding their weekend show. Their first mission is to fly the area and to become familiar with its landmarks. San Franciscans will be thrilled by their precision maneuvering as they reconnoiter the area. Zoom, right past the Transamerica Tower!

Friday is practice day. For me, practice day is often better than the Saturday or Sunday performances. You've got the Blues without the airshow crowds. The best place to view both arrival day and practice day is Fisherman's Wharf or Pier 59. Get there early to find an unobstructed view. Bring your camera. A picture of the Blues sweeping down over the Golden Gate Bridge and screaming low over the Bay with Sausalito in the background belongs in your scrapbook.

Saturday is a spectacular day. The airshow starts at noon. Find yourself a nice spot at Golden Gate Park early. Ships of the U.S. Navy will sail from the Pacific under Golden Gate Bridge and into the bay for a 1-hour preair show **Parade of Ships.** The ships will then dock at San Francisco and East Bay piers to open for public tours on Sunday only.

In 2003, the **San Francisco Maritime National Park Association** hosted a Fleet Week viewing party and reception atop its spectacular art deco Maritime Museum building. Members and the general public were invited to view the spectacular Blue Angels airshow from the decks of the ship-shaped Maritime Museum building. This is one of the premier viewing areas. If they are doing it the year you visit, plan to attend. Check their Website for further information and pricing (http://www.maritime.org).

If the Maritime Museum is not available, you're not out of luck. The airshow can also be well seen from bay-view rooms in the Fairmont Tower or the Mark Hopkins. A really great spot is the Top of the Mark revolving restaurant. All three of these options are high-priced and hard to get a reservation for during the show. Try! You may get lucky and the view will be well worth your trouble. The memory of the airshow will linger long after the misery of the price has past!

Dates

Early October each year

Hours

Saturday noon–5:00 p.m.
Sunday noon–5:00 p.m.

Admission Cost

Free

Contact Information

http://www.fleetweek.com/sf/

G/A Airport Serving the Event

Oakland, CA (Metropolitan Oakland International—OAK)
Phone: (510) 577-4000
Tower: 118.3
ATIS: (510) 383-9514

There are many GA airports in the bay area. None are closer and more convenient than Oakland. You can catch a shuttle bus to the nearest BART station and the trip to San Francisco. You really don't want to drive into San Francisco during Fleet Week. It will be extra crowded, and parking spaces will be hard to find and priced to match the overcapacity demand.

FBO Name/Phone Number

Sierra Jet Service
(510) 568-6100

Ground Transportation

BART

Best Place to Stay $$$$$

The Fairmont San Francisco
(415) 772-5000

This is the place. The Fairmont has been here for a really long time. It is actually two hotels, the Towers and the original Grand Dame of San Francisco. The Fairmont was one of only a handful of large buildings to survive the Great Earthquake of 1906. It has been refurbished to its original splendor many times since then. It is spectacular but don't stay in it. For this trip and this trip only you want a room in the Fairmont Towers. Spring for a suite on the northwest corner. You'll be thrilled with the view. Without an air show it is amazing; with one it is indescribable.

The Fairmont has been my address when in San Francisco for many years. Make it yours for this airshow.

Best Aviation Attraction

#38

The Florida Air Museum at Sun 'n Fun

The **Florida Air Museum at Sun 'n Fun** and Aviation Center is located on a 40-acre campus area adjacent to the fly-in convention site. The 20,000-square-foot facility houses the museum's current collection of aircraft and memorabilia. It contains a small library shelving 4000 aviation books and currently houses a collection of 35 experimental and homebuilt aircraft.

The groundbreaking ceremony for phase I of the Florida Air Museum at Sun 'n Fun and Aviation Center was held in November, 1991, and its doors officially opened for the first time April 6, 1992.

When completed the **Florida Air Museum at Sun 'n Fun** will consist of 100,000 square feet of display, restoration, educational, and administrative spaces. A four-story, central core facility will be joined with the museum exhibit wing to house a variety of aviation displays and aviation artifacts with emphasis on sport aviation. An educational complex consisting of theater, classrooms, and research facilities is planned, along with an expanded library and conference center.

There are lots of reasons to come here, certainly the display of homebuilt experimental and ultralight aircraft is compelling. There are some production planes of period importance here as well. The collection of aircraft is at this time modest. However, if you time your visit to take advantage of one of the special events conducted throughout the year, your visit will be productive and you will be delighted. For example, Richard VanGrunsven, founder and president of Van's Aircraft, spoke at the **Florida Air Museum at Sun 'n Fun** in January 2004. Check the schedule to see who will be speaking during your visit.

I am often asked if it is a good idea to plan your visit to the museum during Sun 'n Fun. My answer is no. There is so much to see at the convention that you wouldn't want to use any of your time at the museum unless you are planning to spend several days at Sun 'n Fun. In that case, consider that the museum will be very crowded and you won't have an opportunity to really admire and study the exhibits to the extent you would prefer.

This is a great museum because it is in Central Florida very near to the theme parks of Orlando and the beaches of Sarasota. When you plan to bring your family to Disneyworld, visit this museum also. It is not worth making a trip from Los Angeles or New York to see. This is not the case for its big brother museum at Oshkosh, which is worth seeing no matter how far you have to travel.

Dates

Open daily
Closed Christmas Day

Hours

Monday–Friday	9:00 a.m.–5:00 p.m.
Saturday	10:00 a.m.–4:00 p.m.
Sunday	Noon–4:00 p.m.

Admission Cost

Adult	$8.00
Senior	$6.00
Students (8–12)	$4.00
Children (7 & under)	Free

Contact Information

G/A Airport Serving the Event

Lakeland, FL (Lakeland Linder Regional—LAL)
(863) 648-3299

Ground Transportation

Rental car

Best Place to Stay $$$$

The Ritz-Carlton Orlando, Grande Lakes
4012 Central Florida Parkway
Orlando, FL 32837
(407) 206-2400
(407) 206-2401 (fax)
http://www.ritzcarlton.com

Here's the best news about Lakeland, Florida. It is only 40 miles from Orlando with its world-class resorts and unequaled theme park. This hotel and spa opened in 2003. From the moment you walk in you know you're in a Ritz. Reward your spouse with a spa package or a day on the golf course. It is the best way to say thanks for hanging out at an airshow with me! For a few dollars less you can stay in the attached Marriott, which is also brand new! Either way you're in for a treat.

Best Aviation Attraction

#39

Florida International Air Show

The U.S. Air Force's precision flying team, **The Thunderbirds,** made their fifth appearance at the 2004 edition of **The Florida International Air Show.** Punta Gorda was the kickoff event for **The Thunderbirds** 2004 air show season. **The Thunderbirds** began their second half century of performing above the enthusiastic southwest Florida crowds. The team flies the Lockheed Martin F-16C Fighting Falcon. This all-weather fighter reaches speeds greater than 1300 mph and altitudes above 50,000 feet. The Fighting Falcon is a true multirole aircraft, and it has an unequaled record in actual air-to-air combat. Each year the principal act changes. Sometimes it is the Blue Angels and sometimes it's the Thunderbirds. Both are AWESOME!

Several other flying performers round out each year's show in addition to the numerous static displays. The show is always big and crowded. My suggestion is that you forget the Saturday and Sunday air show and focus on Friday's practice day. All of the acts will be given a practice slot and they will all use it. Normally, the principal act will fly in the morning and again in the afternoon. Park along the fence line of the airport, unload your cooler and chairs and enjoy this near-private show. If you do go to either a Saturday or Sunday performance, remember that you will not be allowed to enter with a chair or cooler. Bummer! Practice should start at 11:00 a.m. and go to 4:00 p.m.

But what will you do on Saturday and Sunday after you've traveled all this way? Simple, go to the beach. Some of the best beaches in the world are within 25 miles of the airport. An air show and a weekend on the beach, during one of Florida's most beautiful seasons are hard to beat. Leave the cold weather of the Midwest behind and get on down here. You'll be glad you did.

Dates

Late March annually (March 27–28, 2004)

Hours

Gates open 8:00 a.m.
Show starts 12:00 p.m.
Show ends 4:30 p.m.

Admission Cost

Adult	$15.00
Children	$5.00

Contact Information

Florida International Air Show
(941) 255-4000

G/A Airport Serving the Event

Punta Gorda, FL (Charlotte County Airport—PGD)
Phone: (941) 639-4119
CTAF-Unicom: 122.7
ASOS: (941) 639-0076
Runways:

3–21	6580′ × 150′
15–33	4743′ × 150′
9–27	4591′ × 150′

PGD will be closed for the air show the following date and times (all times local):

Friday	1100 hours to 1600 hours
Friday	1700 hours to 2200 hours
Saturday	1200 hours to 1700 hours
Sunday	1200 hours to 1700 hours

Aircraft parking will be on runway 9–27. The CAP will provide transportation to and from your aircraft.

FBO Name/Phone Number

Charlotte County Airport Authority
(941) 639-4119
http://www.heraldtribune.com/apps/pbcs.dll/section?CATEGORY=AIRSHOW

Ground Transportation

Rental car

Best Place to Stay $$$$

Sundial Beach Resort
Sanibel Island, FL
(239) 481-3636

Best Place to Stay $$

Courtyard—Fort Myers Cape Coral
4455 Metro Parkway
Fort Myers, FL 33916
(239) 275-8600

Best Aviation Attraction

#40

Flying Leatherneck Aviation Museum

Flying Leatherneck Aviation Museum is located aboard the Marine Corps Air Station, Miramar in California. Enter the base through the north gate and tell the sentry that you want to visit the museum. Be certain to have a photo ID for every person in your party who is 16 years of age or older. The driver must have a valid operator's license, vehicle registration papers, and proof of insurance if the car is to be driven on the base.

Since September 11, 2001, access to the **Flying Leatherneck Aviation Museum** has become more difficult. If the sentry turns you away, call the museum for assistance. You can also call 693-1723 (FOUNDATION) ahead of your visit to arrange for an escort. An escort may not be available at all times because of manpower. There may soon be a private gate for the museum.

This is the only museum in the world that displays the tools of the Marine Corps aviator. Many of the planes that we have all seen in movies and television shows about World War II, Korea, and Vietnam are here. Eighty-eight Marine Aces flew the Chance Vought F4U-5N Corsair. This is the most memorable U.S. aircraft of World War II's Pacific theater. The Black Sheep squadron of "Pappy" Boyton flew this zero eater.

I was a Vietnam era mud-Marine and the three aircraft that meant the most to me are all here. Maybe you remember them as well. The Rockwell (North American) OV-10 Bronco, the McDonnell Douglas F-4 Phantom, and the Bell AH-1J Sea Cobra. These guys came around when you needed a little help, real quick!

All of the displayed aircraft are in more than perfectly restored pristine condition, hanging from the ceiling in a climate-controlled facility. Somehow they seem more real in this setting.

At least half the reason to come here is the opportunity to drive around this base because just a few years ago it was the Navy base that hosted **Fightertown USA** and **TopGun.** The best of the best came here to train in those days. It is a privilege just to be here.

Dates

Open daily

Closed Sundays and holidays

Hours

Monday–Saturday 9:00 a.m. to 3:00 p.m.

Admission Cost

Free

Contact Information

Flying Leatherneck Aviation Museum
MCAS Miramar
P.O. Box 45316
San Diego, CA 92145
(858) 693 1723
flhf@flyingleathernecks.org
http://www.flyingleathernecks.org/

G/A Airport Serving the Event

Carlsbad, CA (McClellan-Palomar—CRQ)
Phone: (760) 431-4646
CTAF-Unicom: 118.6
AWOS: (760) 930 0864
Runway: 6–24 4900′ × 150′

FBO Name/Phone Number

Western Flight
(760) 438-6800

Ground Transportation

Rental car

Best Place to Stay $$$$$

La Costa Resort and Spa
Costa Del Mar Road
Carlsbad, CA 92009
(800) 854-5000
http://www.lacosta.com

Best Aviation Attraction

#41

Get a New Rating

The only requirement the FAA places on you to keep your license active is the Biennial Flight Review (BFR). I have always objected to it because it doesn't follow a set syllabus nor is there any criterion for passing it. Theoretically it is not a pass/fail event. The theory is lost in the real world because what is required is a sign-off which will come only at the discretion of the flight instructor you happen to be working with. The flight review can go on for 1 or 2 hours or 1 or 2 days. It is totally up to the instructor you chose. Choose poorly and your wallet can be needlessly drained.

Fortunately, there is a more productive alternative with a more predictable outcome. Get yourself a new rating and it counts as the BFR. The benefits are enormous particularly for owner pilots. Often your insurance rates go down as your competency as a pilot goes up. Makes sense doesn't it?

Most of us start off with a private pilot's license and a single-engine land rating. Be careful when you select the next step: it can be a showstopper. The next logical rating to pick up is the instrument rating. With it you can have a more controllable schedule when you are using your aircraft as a traveling machine. There is nothing worse than planning your family's Bahamian vacation 3 months in advance and finding the weather has turned to "VFR not recommended" at the moment you need to take off. My family and I once left our airplane in Atlanta and flew back with Delta because of fog and low ceilings. You know the condition: runway visibility 1 mile in fog, cloud bottoms at 400, tops at 1500, and clear blue skies from there on up to the moon. The problem is you can't take off VFR with that weather report, and you can't climb through the 700 feet of scud to get to the VFR on top conditions. That's when your brain will start screaming instrument ticket.

Be warned! The instrument rating is the hardest rating in aviation to get. Only 10 percent of the private pilots in the United States have an instrument rating, 10 percent! The reason is it is powerfully difficult to get and the expense is high. It takes as many flight hours for your instrument rating as it does for your private pilot's license. Most folks take much longer than 1 year to finish it once they have started. Recognizing this fact the FAA allows you 24 months following your successful passage of the written exam to pass the checkride.

The problem in pursuing the coveted IFR ticket is that discouragement can easily set in. It took me 18 months, three flight instructors, multitraining aids and two checkrides to finally get it in my wallet. I blew the first checkride which also is not unusual. This is a tough act.

My experience led me to the following conclusion: computer-aided flight instruction is key to passing the written for any rating requiring a written examination. John and Martha King have got the best stuff out there. Use them, you'll be glad. *Sporty's* has a wonderful *free* online IFR study guide and IFR practice test. It was really helpful to me and I am grateful for it. I wasn't helped by the *Gleim* manuals at all, I actually hated them. Many people swear by them, I swear at them! Some folks go to a weekend cram course like the ones given by American Flyers. Be careful with this approach particularly if the training city you will be going to is some distance from your home. You are encouraged to wait a day after taking the class to take the test. If you don't pass, you can come back again and again and again. Not a bad offer if they are close, but waiting that extra day for testing is a hurt. Plus you really do want to know the material, not merely pass the test. My advice is to buy the King IFR course. Take it all the way through. Then go up on the *Sporty's* site and take the test over and over again until you consistently understand the material and score over 70 percent. Then sign up to take the test at one of the many computerized test facilities across the country. The completion certificate from the King course is sufficient for the instructor sign-off. Take the test and pass it. Now engage an instructor and start flying. Plan on at least 4 hours per week. I know this seems like a lot of flight time but it will still take you at least 2½ months to fly off the 40-hour requirement. When you are not with your flight instructor be at home on your *Microsoft Flight Simulator 2004* equipped computer. Practice, practice, practice. It is a hard rating but you can do it if you put your mind to it, and it really will make your flying more practical.

OK so the IFR is the big tuna and it will take almost the entire 24 months between BFRs to get it. There are some other fun, quick routes. How about single-engine *water?* You can knock this rating off in a weekend at several spots across the country. You will likely never use it again, but it is a hoot to get and it counts for the BFR. If in doubt, always ask before you sign up and check it out with the FAA FSDO and your insurance agent. Next pick up a multiengine rating. This is a semiquickie. It takes about 10 hours. When you take the checkride be sure to shoot one single-engine IFR approach so you can get not only a multiengine but a multiengine IFR rating. The add-on is a piece of cake.

Any power pilot should spend some time in a glider and pick up a glider rating. You will come away a much better pilot. These guys always nail approaches perfectly; there is no throttle to pull or push to compensate for poor approach planning.

The most meaningful step of all is to trade in your private pilot's license for a commercial pilot's license. This is an easy step compared to the IFR test you've already been through by this point. The air transport pilot rating isn't far away by now. You'll probably never become an airline pilot and probably don't want to fly for an occupation anyway. It is all about being a better, safer pilot, but wouldn't the pro card look great in your wallet.

Once the ratings have all been obtained what else can you aspire to? Type ratings of course! Get checked out in everything with wings you can find. Be the best you can be pursuing your hobby to the max!

Best Aviation Attraction

#42

Get Inverted

It is said that the fruit never falls far from the tree and it is true. **Harvey & Rihn Aviation** produces world champions. In 2003 and again in 2004 Debby Rihn-Harvey's superior piloting skills have earned her the title of *Women's U.S. National Aerobatic Unlimited Champion.* For more than 20 years H&R has been a world-renowned leader in aerobatic instruction. Whether you aspire to be a competition-level pilot or you just want to increase your level of confidence, H&R can provide the appropriate level of aerobatic training to meet your needs. It's legendary for its spin training course.

Most pilots come to H&R in Texas not to be aerobatic champions but to become better pilots and to that end the **Safety Proficiency Course** was developed. It is designed for any pilot who would like to be in complete control of his aircraft in unexpected situations. The course includes spins, rolls, and recovery from very unusual attitudes. The time required to complete the course varies with flight experience and will take 2 to 3 hours. If your interest is greater, perhaps the **Situational Awareness Course** is best suited for you. It aims to enhance a pilot's awareness of the overall flight situation and allow precise control of the airplane around all of its axis. Therefore, the mission at hand (patrol, search and rescue, etc.) can take priority with safe and skillful flying being second nature. The course is divided into five flights, totaling 5 hours, that progress from simple proficiency maneuvers to more demanding unusual and inverted flight. Spin training is involved in both of these courses. You will leave with a greater confidence in your piloting ability knowing you can handle your airplane in any situation or attitude.

The cost is much less than you might imagine. The fee for an aerobatic instructor is $36 per hour. While you may own an aircraft it is probably not certified for aerobatics. H&R has several planes available. You may rent a Citabria for $72 per hour wet or a Decathlon for $108 per hour also wet. Either of these planes are suitable for aerobatic training. I am not certain why you would choose one over the other but I am certain that experts at H&R know and can advise you appropriately. If you go with the Citabria and avail yourself of the cash discount policy the instructor and the Citabria will cost you $105 per hour wet. At this rate you can walk away as a much better pilot for somewhere between $200 and $500.

You will probably be coming in from out of town so it is good to know that you can tie your ship down, buy fuel, and take your training all at H&R. If you take the Situational Awareness Course, it is probably best to split the course over 2 or more days. This training is intense and 5 hours of aerobatics in 1 day is far too much. With careful planning, you fly in, take the Safety Proficiency Course, and fly out all in the same day. My advice is to fly in the night before and begin your training early the next morning when you are fresh and the air is smooth.

When it comes to aerobatic training you really should come to **Harvey & Rihn Aviation.** They know their stuff!

Contact Information

Harvey & Rihn Aviation
101 Airport Boulevard
La Porte, TX 77571
(281) 471-1675
(281) 471-6646 (fax)
info@harveyrihn.com
www.harveyrihn.com

G/A Airport Serving the Event

La Porte, TX (La Porte Municipal Airport—T41)
CTAF-Unicom: 122.7
AWOS: 713-847-1462
Runway: 12–30 4165′ × 75′

FBO Name/Phone Number

Harvey & Rihn Aviation
(281) 471-1675

Ground Transportation

Rental car

Best Place to Stay $$$

Residence Inn—Clear Lake
525 Bay Area Boulevard
Houston, TX 77058
(281) 486-2424

Best Place to Stay $$$

Hilton—NASA
3000 NASA Road One
Houston, TX 77058-4322
(281) 333-9300

Best Aviation Attraction

#43

Grand Canyon Discovery Flight

The most magnificent natural formation on the earth is the Grand Canyon. People come from all over the world to see it. That is the problem—seeing it! Fortunately those of us who love airplanes and adventure have a real treat in store. You can tour the Grand Canyon by airplane or helicopter. It is no longer possible to do this with your own ship as flight restrictions keep you too high to really enjoy the Canyon. Also, you are not a trained tour guide who really knows this part of the world and what to see. Further, if you're enjoying the view whose flying the plane? If you're not enjoying the view but concentrating on flying the plane, why are you here?

The answer is to use one of the many air tour operators who are permitted to fly the canyon in fixed- and rotary-wing aircraft. In my opinion the best is Grand Canyon Airlines. They are also the FBO at **Grand Canyon National Park Airport.** These guys have been doing this trip since 1927 and have flown millions of passengers in the last 26 years. Today they use a Twin Otter—Vistaliner rather than a Ford Trimotor.

The tour you'll fly covers the most famous and beautiful parts of the Grand Canyon. Narration is provided digitally in several languages, and is designed to inform and entertain.

Dates

Daily

Hours

Hourly departures based on demand

Admission Cost

Adult $75.00

Child (2–11) $45.00

If you prefer you can go in a Bell helicopter for an additional $100/adult and an additional $110/child. It gets you up close and personal with Mother Nature's handy work.

Contact Information

Grand Canyon Airlines

(866) 2-FLY-GCA

G/A Airport Serving the Event
(Grand Canyon National Park—GCN)
Phone: (520) 638-2446
Tower: 119.0
AWOS: (520) 638 0672
Runway: 3–21 8999′ × 150′

FBO Name/Phone Number
Grand Canyon Airlines
(520) 638-2463
www.grandcanyonairlines.com

Ground Transportation
Tour bus or rental car

Best Place to Stay $$$$:
Best Western Grand Canyon Squire Inn
Highway 64
Grand Canyon, AZ 86023
(928) 638-2681
http://www.grandcanyonsquire.com/

Best Aviation Attraction

#44

Hawaii by Helicopter

Some say that the most beautiful spots on earth are in the Hawaiian Islands. I agree. The problem is that 99 percent of tourists that go to Hawaii never see them. Most are totally inaccessible.

Most of us have seen helicopters flying. To the ground-bound they are often nothing more than noisy nuisances. Helicopters are about utility. If you want to get somewhere that doesn't have an airport quickly or lift something heavy when there isn't a crane, then, my friend, those noisy nuisances become a beautiful answer. If you're the Donald trying to get to your Manhattan office or The Parks and Wildlife Department trying to put out a fire in Idaho's back country, you'll want one. They are also the answer to seeing the undiscovered Hawaii!

It doesn't really matter which Island you're on, there is a heli-tour operator close. Be careful, quality control of personnel and equipment varies widely. In my experience the one to use is **Blue Hawaiian Helicopters.**

My favorite island in the chain is Maui. It has everything I like, great golf, terrific beaches, a good restaurant, a small town for walking around and shopping, a tall mountain (volcano), and a remote village (Hana). **Blue Hawaiian Helicopters** has a 30-minute flight which is affordable at $125 and wonderful. You'll explore deep, meandering valleys set in the rainforest of the ancient West Maui Mountains. Fly over, around, and next to the knife-edged ridges with mist-shrouded peaks that separate the valleys that give Maui its nickname, *The Valley Isle.* Waterfalls cascade down from towering cliffs into the streams running along the rainforest floor. At one point there are 17 that flow into each other. Another has a 1100-foot drop. Every tour is unique as changes in wind and weather determine which sites are best for your flight. The warm tropical climate and year-round trade winds add a dynamic element to the beauty of the islands. The waterfalls and rainbows can be even more spectacular.

Plan to see Hawaii by helicopter so that you don't miss the 90 percent of Hawaii that cannot be seen without one.

Dates

All the time

Hours

Ready when you are but book in advance.

Admission Cost

$125.00 per person for the 30-minute flight.

Want more? Spend more!

Contact Information

Blue Hawaiian Helicopters
105 Kahului Heliport
Kahului, HI 96732
(800) 745-2583
(808) 871-8844
info@bluehawaiian.com
http://www.bluehawaiian.com

Best Place to Stay $$$$$

Kapalua Bay Hotel and Ocean Villas
One Bay Drive
Kapalua
Island of Maui, HI 96761
(808) 669-5656

Best Place to Stay $$$$$

Fairmont Kea Lani
4100 Wailea Alanui Drive
Wailea
Island of Maui, HI 96761
(808) 875-4100

Best Aviation Attraction

#45

Hill Aerospace Museum

Hill Aerospace Museum is located at Hill Air Force Base which is about 5 miles from Ogden, Utah. It annually hosts 200,000 visitors. They come to see exhibits of over 70 military aircraft, missiles, and aerospace vehicles on the grounds and inside the **Major General Rex A. Hadley Gallery** and the **Lindquist-Stewart Fighter Gallery.** The museum collection includes a wide variety of ordnance, an assortment of aerospace ground equipment, military vehicles, uniforms, and thousands of other historical artifacts.

Plane Talk is a series of free lectures held on Saturdays at 1:00 p.m. in the museum classroom during the fall and winter of each year. Test pilot Scott Crossfield, Tuskegee Airman Col. Herbert Carter, F-4 pilot Brig. Gen. Robin Olds, and Gen. Robert L. Scott (author of "God Is My Copilot") have all spoken here. In early 2004 Chief Warrant Officer Richard Burt discussed his experiences as a gunner on a B-24. He was shot down on November 17, 1944, on only his fourth mission. Stalag Luft IV then became his home until the end of the war. Colonel James Sullivan (USAF, retired) flew both the Lockheed A-12 and SR-71 Blackbird. Sullivan set a speed record which has never been equaled by flying from New York to London in less than 2 hours. If you could be in the presence of these men, what would you ask them?

Think about it, 200,000 people a year choose to come here. Ask yourself why. The reason is the vast collection, and it really is vast. Air power from *every* era is on display including a B-1B. They also display an impressive amount of the equipment that keeps these machines in the air and the armament that they carry. Ever seen an atom bomb? They've got a couple here.

If you are touring the Rockies for any reason (skiing is a good one), make a stop here. Check the schedule first and see who's giving a *Plane Talk* lecture. Time your visit to lend an ear if you can.

Dates

Open daily

Closed: New Year's Day, Thanksgiving Day, and Christmas Day

Hours

9:00 a.m.–4:30 p.m.

Admission Cost

Free

Contact Information

Hill Aerospace Museum
7961 Wardleigh Rd.
Hill AFB, UT 84056-5842
(801) 777-6868
http://www.hill.af.mil/museum/

G/A Airport Serving the Event

Ogden, UT (Ogden-Hinckley Airport—OGD)
Phone: (801) 629-8251
Tower: 118.7
AWOS: (801) 622-5600
Runways:

3–21	8103′ × 150′
7–25	5600′ × 150′
16–34	5195′ × 150′

FBO Name/Phone Number

Ogden Jet Center
(801) 392-7532

Ground Transportation

Rental car

Best Place to Stay $$$$

Marriott
247 24th Street
Ogden, UT 84401
(801) 627-1190

Best Place to Stay $$$

High Country Inn
1335 West 12th Street
Ogden, Utah 84404
(801) 394-9474
(801) 392-6589 (fax)

#46

Horseshoe Bay Resort

Here you'll find a playground created for those who have earned and deserve the finer things of life. Nestled along the shores of Lake LBJ, **Horseshoe Bay Resort** is one of the nation's most luxurious, international caliber, destination resorts, featuring an unequalled combination of amenities, services, and facilities. Meticulously designed and crafted, this resort spans over 6500 acres in the center of the Texas hill country.

Horseshoe Bay Resort offers three pool options and a private beach on the constant-level Lake LBJ. Personal watercraft, pontoon boats, paddle boats, ski and jet boats are available for watersport enthusiasts. It is also home to three Robert Trent Jones, Sr., golf courses, each distinctively different in style and degree of difficulty.

If tennis is your game you're in luck, there are eight outdoor Laykold-surfaced tennis courts and six outdoor clay courts. The Racquet Club tennis facility is surrounded by the exquisite Oriental Water Gardens that contain fountains, waterfalls, and fitness trails.

The staff will see to your every need. Their goal is that each of the guests departs with memories to last a lifetime!

It is a world-class resort with a truly world-class private airport. The terminal building houses a luxury lounge and pilot ready room. The airport is lighted for night operation, and features full facilities for 24-hour prop and jet maintenance, communication, and navigation. The 6000-foot runway can accommodate aircraft up to DC-9 jets. The best part about this airport is its security. No one may land here without prior permission. You and your passengers are free from unwanted intrusion. The airport is a fitting gateway to the sanctuary offered by this world-class facility.

Dates

Year round

Contact Information:

Horseshoe Bay Resort
(830) 598-2511
frontdesk@horseshoebaytexas.com

http://www.horseshoebaytexas.com/

G/A Airport Serving the Event

Horseshoe Bay, TX (Horseshoe Bay Airpark—4XS7)
E-mail: airpark@horseshoebaytexas.com
Phone: (830) 598-6386
CTAF-Unicom: 122.8
AWOS: (512) 756-7277
Runway: 17–35 6000′ × 100′

Because they understand that every second of leisure time is precious, they've built a private airport. It is one of the finest in the United States. You may only land here with prior permission.

FBO Name/Phone Number

Horseshoe Bay Airpark
(830) 598-6386

Ground Transportation

None required

#47

Imperial War Museum Duxford

As the European center of aviation history, the **Imperial War Museum Duxford** in the United Kingdom retains its wartime bearing. Over 400,000 visitors come annually to see the 180 historic aircraft on display: biplanes, Spitfires, a Concorde, and jets from the Gulf War. The site houses outstanding exhibits and regularly schedules air shows. The combination creates a unique museum where history really is in the air.

The **American Air Museum** in Britain stands as a memorial to the 30,000 Americans who died flying from the United Kingdom during the Second World War. This museum houses a unique collection of historic American combat aircraft including the B-17 Flying Fortress and the unstoppable B-24 Liberator. Many aircraft from the collection are suspended from the ceiling as if in flight. This museum is but one section of the Imperial War Museum at Duxford.

The aerodrome at Duxford was built during the First World War and was one of the earliest Royal Air Force stations. RAF Duxford became No.2 Flying Training School in 1920, equipped with the Avro 504, the DH9A, and the Bristol Fighter. The period of intense air fighting during World War II has become known as the *Battle of Britain*. Duxford earned its true glory during this period. By 1941 it would become a key launch point for the allied bombing campaign of Europe, central to the liberation of Europe, which was to come.

The Ministry of Defense declared its intention to dispose of the airfield in 1969. The Imperial War Museum had been looking for a suitable site for the storage, restoration, and eventual display of exhibits too large for its headquarters in London and obtained permission to use the airfield for this purpose.

When you come here you will enter history. You may be able to join forces with a companion who served here during World War II. You can find them touring the museum and reliving a part of their youth which they certainly wished not to live and now are driven to never forget. Introduce yourself, thank this new-found friend for his contribution to our society and ask him a question or two. Soon he'll tell you stories about this place and the air-war that was fought above and from it that are certain to have been missed by every historical writing. It is rare to visit a battle field in the company of a warrior who fought on it. Do not let this opportunity pass you by.

If he was a crew member of a B-24, stand beside it as you talk. Get him to tell you what it was like, the sounds, the smells, the fear, and the glory that became his companions during his long missions across the channel. What did it feel like? Was it cold? Did the plane jump off the runway or struggle to find its way into the air? What was it like to look down on the fields of France that were crawling with an enemy, his enemy, an enemy intent on shortening his mission and his life? Was he shot down? How did he survive?

The museum at Duxford is a monument to his history and yours and mine, and to the aircraft that protected us all. The frail tools that we call fighters and bombers were hard enough to fly and keep in the air without being the object of angry enemies. Imagine riding in a B-17 at 25,000 feet. Remember they were not pressurized or heated and a substantial portion of each side of the fuselage was left open for the waist machine gunners to practice their trade. It was cold, very cold and very drafty. Ice formed inside as well as out. Men died because their moist breath froze inside their oxygen mask and blocked the flow of the essential gas that meant life.

Listen to the stories, read the history, touch the airplanes, and imagine the terror!

Museum Dates

Open daily

Closed December 24, 25, and 26

Museum Hours

Winter	10 a.m. to 4 p.m.
Summer	10 a.m. to 6 p.m.

2004 Airshow Schedule

D-Day Anniversary Show

Sunday 6 June

Flying Legends Air Show

Saturday 10 and Sunday 11 July

Duxford 2004 Air Show

Saturday 4 and Sunday 5 September

Autumn Air Show

Sunday 10 October

Admission Cost

Adult	£8.50
Seniors (person aged 60 years or more)	£6.50
Students	£4.50
Children (15 and under)	Free

Contact Information

Imperial War Museum Duxford

Cambridgeshire

CB2 4QR

United Kimgdom

01223 835 000

01223 837 267 (fax)

duxford@iwm.org.uk

http://www.iwm.org.uk/duxford/index.htm

G/A Airport Serving the Event

You may land at Duxford with prior permission. Telephone 01223 833376 to obtain permission.

Runways

06/24	1503m × 45m Asphalt
06/24	890m × 30m Grass

Communications

Duxford Information	122.075
Cambridge Approach	123.6
Stansted Radar	120.625

FBO Name/Phone Number

The Imperial War Museum Duxford

Ground Transportation

Rental car. A free bus service runs from Cambridge.

Best Place to Stay $$$$$

Hotel Felix Cambridge
Whitehouse Lane
Huntingdon Road
Cambridge CB3 0LX
United Kingdom
01223 277977
01223 277973 (fax)
help@hotelfelix.co.uk
www.hotelfelix.co.uk

Best Place to Stay $$$

Duxford Lodge Hotel
Ickleton Road
Duxford
Cambs CB2 4PP
United Kingdom
01223 836444
01223 832271 (fax)
duxford@btclick.com
www.duxfordlodgehotel.co.uk

#48

International Trade Fair for General Aviation—Friedrichshafen

Fifty thousand people are expected to attend the 15th **International Trade Fair for General Aviation,** (**AERO**) in Friedrichshafen, Germany. It is the largest general aviation trade show in Europe. More than 700 exhibitors from 30 countries will present their products to a multinational audience. Seventy percent of the attendees hold a pilot's license and have a real interest in the products being shown, talked about, and flown. The 4-day event will take place for only the second time on the new exhibition grounds, which are immediately adjacent to the airport.

According to the trade fair's management, exhibitors will showcase the entire spectrum available to general aviation, ranging from business jets, gliders, and ultralights to onboard electronics, maintenance, and accessories. European general aviation companies save their most important announcements for this show. The exciting new twin diesel from Diamond was first shown here as was its single-engine jet.

This is not only a trade show of static display but an air show as well. On the first 2 days, Thursday and Friday, manufacturers will be flying their latest offerings in skies above Friedrichshafen. After landing, these craft will be taxied to the show's static display area for visitor inspection.

On Saturday and Sunday the best European flyers will put on a show using the historic aircraft they have brought to the show. In 2005, as in 2003, the crowd will be thrilled to see the following airmen and aircraft take to the skies:

Bender, Gert Bücker	131 Jungmann
Bucher, Andreas	D17 Staggerwing
Eichhorn, Walter and Toni	L-29 "Delfin," 2 × T 6
Marwig Herzog	WACO YKS-T
Mathis, Marc	YAK 3U
Niebergall, Ralf	SF-260
Storch von, Thomas	Slepcev Storch
Benning, Uli	CAP Aviation
Scheid, Arthur	PT 17 "Kaydet" Stearman
Ultimate High Power Aerobatics Academy	Extra 300
Vogelsang, Max	P51 Mustang
Fliegerrevue Airshow-Team	4 Zlins

Extra, Walter Extra 400
Schroth, Klaus Extra 300

AERO visitors will be able to also visit the zeppelin hangar located on the exhibition grounds, during all four show days. The **Airship Hall** is open during the entire show. At the information kiosk visitors will be able to find out more about airship construction and the zeppelin's role in contemporary aviation.

Show Dates

Convention April 21–25, 2005
Airshow April 23–24, 2005
Exhibitor flight demonstrations April 21–22, 2005

Show Hours

Daily 9 a.m. to 6 p.m.
Exhibitor flight demonstrations 1:00 p.m. to 3:00 p.m.

Contact Information

Customer Service—AERO
49 07541 / 708-404
49 07541 / 708-2404 (fax)
besucher@messe-fn.de

Project Management—AERO
49 7541 / 708-361
49 7541 / 708-2361 (fax)
aero@messe-fn.de
http://www.messe-fn.de/fairs/aero/index.php3

G/A Airport Serving the Event

Flughafen Friedrichshafen GmbH
49 (0) 7541 / 284-01
49 (0) 7541 / 284-119 (fax)
info@fly-away.de
http://ww2.fly-away.de/en/

Ground Transportation

AVIS
Reinbold GmbH
Agentur der Avis (Autovermietung GmbH & Co. KG)
49 (0) 7541-930700
49 (0) 7541-930705 (fax)
http://www.avis.de

Getting Here from the United States

Without a doubt, the most convenient and interesting access to this museum is through the Zurich International airport. While there are airlines offering connecting flights to Friedrichshafen, don't take one. Instead go directly to the train station which lies just under the airport terminal. Take the train to Romanshorn. From Romanshorn you travel across beautiful Lake Konstancevia on a ferry to Friedrichshafen. The ferry terminal is a short walk from the town center. During the air show there will be a shuttle to take you directly to the fair grounds. Alternatively, there are taxis.

Best Place to Stay $$$

Best Western Hotel Goldenes Rad
Karlstrasse 43
Friedrichshafen, 88045
Germany
0049-(0)7541-28 50
0049-(0)7541-285 285 (fax)

The hotel is located exactly in the town center and is a short walk to the lakefront, the museum, and transportation facilities.

#49

Intrepid Sea-Air-Space Museum

On a beautiful spring day of April 26, 1943, the fourth ship to be christened, *USS Intrepid,* was launched. She was commissioned August 16 of the same year. Captain Thomas L. Sprague was given command. Her service throughout World War II was very impressive.

Following the war *Intrepid* was decommissioned as were many, many ships of the line. In 1952 the *Intrepid* was brought out of mothballs. On October 13, 1954, she entered the jet age and became the first carrier to launch aircraft with American-built steam catapults.

By 1966 *Intrepid* entered battle once again with the Pacific Fleet off Vietnam. Skyhawks and A-1 Skyraiders operated off her decks with launch intervals of less than 30 seconds. *Intrepid* returned to Norfolk and was decommissioned for the final time on March 15, 1974.

A campaign led by the Intrepid Museum Foundation saved the carrier from the scrap pile and established her as a floating museum. It opened in New York City in August 1982. In 1986, *Intrepid* was officially designated as a National Historic Landmark.

The *Intrepid* is the most unique aviation attraction in the world. Berthed on the Hudson River in Manhattan, she contains countless static displays of famous aircraft and educational exhibits. Her berth is shared with the destroyer *USS Edson* and the submarine *USS Growler.* The Intrepid's flight deck is the focal point of any visit. Here you will find aircraft that never flew off a carrier. The Blackbird and the Concorde are the largest examples. Visitors who could never afford supersonic flight across the Atlantic may now tour the Concorde for $14. You may even go to the flight deck and sit in the Captain's chair.

The more adventurous visitors can grab a ride in the A-6 Cockpit Simulator and visit the Virtual Flight Zone. Plan to spend at least a day aboard the *Intrepid.*

Dates

Open everyday except Thanksgiving and Christmas

Hours

Spring/Summer Schedule (April 1–September 28)

Weekdays	10:00–5:00
Saturday, Sunday, and Holidays	10:00–6:00

Fall/Winter Schedule (October 1–March 31)

Mondays	Closed
Tuesday–Sunday	10:00–5:00

Admission Cost

Adults	$14.00
Veterans, U.S. Reserve	$10.00
Seniors	$10.00
College students	$10.00
Students (12–17 years)	$10.00
Children (ages 6–11)	$ 7.00
Children (ages 2–5)	$ 2.00
Children (under 2)	Free

Website

www.intrepidmusuem.org

Contact Information

Intrepid Sea-Air-Space Museum
Pier 86
12th Avenue and 46th Street
New York, NY 10036
(212) 245-0072
info@intrepidmusuem.org

G/A Airport Serving the Event

Teterboro, NJ (Teterboro Airport—TEB)
Phone: (201) 288-1775
Website: www.teb.com
Tower: 119.5
ASOS: 201/393-0855
Runways:

1–19	7000′ × 150′ grooved asphalt
6–24	6013′ × 150′ grooved asphalt

FBO Name/Phone Number

Atlantic Aviation
(201) 288-1740
First Aviation Services
(201) 288-3555
Jet Aviation
(201) 462-4000
Million Air
(201) 288-5040
Signature Flight Support
(201) 288-1880

Ground Transportation

Taking a car into Manhattan is no pleasure. The traffic is world class and parking fees will literally break the bank. Do what New Yorkers do and take a limousine. There are several operators to choose from. Your best bet is to let the FBO make a recommendation and reservation.

Best Place to Stay $$$$$

Hudson Hotel
356 W. 58th Street
New York, NY
(800) 444-4786

Manhattan is filled with luxurious hotels. The Hudson is our pick due to its proximity to the **Intrepid Sea-Air-Space Museum.** It is not only a deluxe hotel, it is also brand new. The decor is captivating. A color-infused escalator, the landscaped outdoor terrace, and the surreal design of the lobby each is in its own way enchanting.

The rooms feature contemporary appointments of warm woods, art, and photographic displays. The Hudson is haute moderne all the way. Westside rooms offer stunning views of the Hudson River. The hotel is only minutes from fine dining, shopping, entertainment, and the Intrepid Museum.

Best Place to Stay $

Holiday Inn Midtown
440 W. 57th Street
New York, NY
(800) 263-9393

The Holiday Inn Midtown 57th Street is New York City's newest midtown location, convenient for major business locations, mass transit, great dining, and sightseeing attractions. It is just down the street from famous theme restaurants like the Hard Rock Cafe, Planet Hollywood, Le Bar Bat, and other attractions, such as Carnegie Hall (three blocks), Lincoln Center (six blocks), and Central Park (3 blocks). Fabulous shopping at Tiffany's, Niketown, and other 57th Street boutiques.

Best Aviation Attraction

#50

Island Hop

Your spouse will love it. The Bahamas are a paradise waiting to satisfy every dream. Here begins a chain of islands that stretches all the way to Venezuela. Once you get hooked on island hopping you'll have years of new adventures in front of you. Any journey begins with a single step and some times the step is a big one. That is certainly true here. Once you're in Florida the closest Bahamian Island is only 56 miles across the water. The problem for most of us is not the 30-minute flight in a Cherokee or Cessna 172. It isn't the distance that gives us pause; it is the water. I suppose this is why castle moats worked so well for so long. Water has a way of holding air breathers at bay.

Here's the way to do it. First latch on to a copy of *Bahamas and Caribbean Pilot's Guide 2004 Edition*. You can get it directly from Pilot Publishing, Inc., by calling the publisher at (800) 521-2120, or you can find it in many pilot shops. Try www.avshop.com or www.sportys.com. Both are well stocked and often have this book. Once you have it, spend several evenings reading, studying, and dreaming.

Next pick up the phone and call Banyan Air Services, Inc. at (954) 491-3170 and ask for the Customer Service Department. You'll be talking to the Number 1 FBO in South Florida and the ultimate solution provider to any Bahamas-bound pilot. They are based at Fort Lauderdale's Executive Airport which is the perfect launching pad for your island adventure. You're going to need some over-water equipment on board which can be purchased or rented at Banyan. These guys know the Bahamas. They can answer any question you may have and will assist you with filling out the required paperwork.

Once you're ready and the trip planning is done, fire up and climb high. You'll be comforted by knowing that you can glide for a while if your engine shuts down. Most of us fly with only one engine and that's a thought. The engine will begin to sound funny the moment you cross the foam line. That's normal. All pilots think they hear strange engine noises when they're over water, and you will too. Less than 1 percent of all aircraft instances involving single-engine aircraft have anything at all to do with the engine. Aircraft engines are very reliable. Next, you will be out of sight of land for less than 15 minutes in the slowest aircraft.

My advice is to make your first landing at Freeport International. It is located on Grand Bahamas Island, which is hard to miss, and is equipped with a VOR and an ILS approach. It is a total of 85 miles from the tarmac at FXE. The trip even in a 172 or Cherokee will take no more than 45 minutes. Runway 8 is 11,000 feet long. You'll have no trouble finding it. Plan your trip for early in the morning. The air will be smooth and the likelihood of thunderstorms is greatly diminished. Do not risk a late afternoon flight. You may run into heavy weather as it pops up in the late afternoon and you *must* land before sundown, not 30 minutes after as in the United States, but at sundown. It is illegal to fly VFR in the Bahamas after official sunset.

Clear customs here and pick up your cruising permit which will allow you to land anywhere within the Bahamas. This will be a piece of cake. The forms are already filled out and you have everything you need. Just do it and be pleasant.

Next, where will you go? Staying right here in Freeport is not a bad option, but if wanderlust has grabbed your heart and you want to get away from the tourists, you'll have plenty of opportunities. Study the *Bahamas and Caribbean Pilot's Guide 2004 Edition* carefully. E-mail, fax, or call the resorts that meet your fancy and make a plan. Many of our friends find happiness at Hawks Nest. Check them out at http://www.hawks-nest.com. It is down on Cat Island and maybe farther than you'll want to go on your first trip.

There are no bad places in the Bahamas!

Dates

Year round. Rates everywhere are 40 percent lower in the summer.

G/A Airport Serving the Event

Fort Lauderdale, FL (Fort Lauderdale Executive—FXE)
Phone: (954) 938-4966
Tower: 120.9
AWOS: (954) 772-2537
Runways:
13–31 4000′ × 100′
8–26 6001′ × 100′

FBO Name/Phone Number

Banyan Air Services, Inc.
(954) 491-3170
info@banyanair.com
www.banyanair.com

#51

Kalamazoo Aviation History Museum

Kalamazoo Aviation History Museum, better known as the **Kalamazoo Air Zoo,** displays many military aircraft. Mostly they are war birds and mostly they are from World War II. They fly their planes often and it is worth the trip to see a P-47 make a low pass or to cover your ears from the noise of the four huge radials of their B-24 taxiing back to its hangar. Each aircraft in the collection has been preserved and restored to airworthy condition. Aircraft from all of America's air wars other than WWI are represented—World War II, the Korean War, the Vietnam War, and the battle over the Persian Gulf. The Air Zoo schedules a "flight of the day" program every afternoon from May through September. But it is so much better than just watching airplanes fly. Weather permitting you can purchase a ride on the Air Zoo's Ford Tri-Motor.

Inside the Air Zoo there is MaxFlight, a full motion flight simulator complete with tilts, turns, wind, and engine noise as well as an F-16 procedure trainer and a KC-135 aircraft cockpit cutaway, all in which visitors are invited to press the pedals, pull the throttle, and push the buttons.

Aviation and art admirers alike can eye the Air Zoo's expansive collection of original photographs, dioramas, scale models, and original oil paintings, illustrating a breadth of experiences and perspectives on wartime, peacetime, military life, and the allure of aviation. If you are a movie buff, you'll be happy here. The museum's 64-seat theater presents classic feature films and award-winning documentaries.

Dates

Open Monday–Saturday
Closed Sunday morning
Closed New Year's Eve, New Year's Day, Easter, Thanksgiving, Christmas Eve, and Christmas Day.

Hours

Summer	9:00 a.m.–6:00 p.m.
Winter	9:00 a.m.–5:00 p.m.

Admission Cost

Adults	$10
Seniors	$8

Children (6–15)	$5
Children (5 and under)	Free

Contact Information

Kalamazoo Aviation History Museum
3101 East Milham Road
Kalamazoo, MI 49002
(269) 382-6555
gservices@airzoo.org
http://www.airzoo.org

G/A Airport Serving the Event

Kalamazoo, MI (Kalamazoo/Battle Creek International Airport—AZO)
Phone: (616) 388-3668
Tower: 118.3
ASOS: (616) 384-5729
Runways:

17–35	6500′ × 150′
9–27	3350′ × 150′
5–23	3390′ × 100′

The Museum is located on the **Kalamazoo/Battle Creek International Airport.** Pilots are encouraged to fly in and park on the museum's ramp. Ground control will give you taxi directions to the Air Zoo.

FBO Name/Phone Number

Kalamazoo Pilots Association
(616) 657-1808

Ground Transportation

Rental car

Best Place to Stay $$$$$

Clarion Hotel
3600 East Cork Street
Kalamazoo MI 49001
(616) 385-3922

Best Place to Stay $$

Holiday Inn—Airport
3522 Sprinkle Road
Kalamazoo MI 49001
(616) 381-7070

Best Aviation Attraction

#52

Kennedy Space Center

(Courtesy of Kennedy Space Center.)

Plan on a full day to explore NASA's launch headquarters, actually 2 days would be better but you can do it in 1 day. My strong advice is that you plan your trip to coincide with a launch, not necessarily a shuttle launch, though that is the most spectacular. Night launches are the best as they light up the sky all the way to the Bahamas.

The Kennedy Space Center is located about 45 miles from Orlando on a huge island which is eight times the size of Manhattan. Everything here is supersized including the rockets. Television does not prepare you for their size. To stand next to the Shuttle replica is a jaw-dropping moment!

Millions of visitors from across the world have made the trek to this hub of technology and discovery, where many of mankind's greatest accomplishments take place.

The Visitor Complex is your entry point to NASA's launch and landing facilities. Here you can experience interactive simulators, live shows and views of massive rockets.

Your trip across NASA's launch headquarters traverses an area that includes towering launch pads, huge rockets, history-making technology, and vast stretches of Florida wildlife. The NASA Up-Close tour is guided by a space program expert who takes you into the operations of Kennedy Space Center.

A highlight of your day will be the 2½ hour **Kennedy Space Center Bus Tour.** It is given daily between 10:00 a.m. and 2:15 p.m. A bus leaves every 15 minutes. There are two stops along the way, the **LC-39 Observation Gantry** and the **Apollo/Saturn V Center.** You'll visit NASA headquarters, and get as close as possible to the Space Shuttle launch pads. You'll also visit KSC's landing facilities, the massive Vehicle Assembly Building, NASA's gigantic Crawler Transporters and the International Space Station Center, where technicians assemble actual components to be launched into space.

Remember that Kennedy Space Center and Cape Canaveral are working space launch facilities. Tours can be altered or closed due to operational requirements. The next launch window for a Shuttle flight is September 12 to October 10, 2004. Atlantis is the scheduled orbiter for the Shuttle programs' return to flight status. If you can possibly be here for the launch you will be very impressed. Countless communication and surveillance satellites are launched from Kennedy Space Center as well. These unmanned launches aboard Atlas, Delta, and Titan rockets are a monthly occurrence.

Additional programs that get you closer to the Space Program are available at additional cost. I highly recommend that you take them. My favorite is the **NASA Up-Close Guided Tour.** Its stops include the A/B Camera Stop, the closest viewing for the Space Shuttle Launch Pads, and the International Space Station Center where actual components are readied for launch into space. Departures for this 90-minute tour are between 10:00 a.m. and 1:50 p.m. daily. Seating is limited and the number of daily departures varies based on operational requirements.

For one of the most memorable meals you have ever had, sign up for **Lunch with an Astronaut.** Seating begins at 12:15 p.m. daily. The program, which includes a briefing by one of NASA's finest, lasts about 90 minutes. Some of America's most famous astronauts including John Glenn and James Lovell have participated in this program.

If you have an extra day and some loose cash I encourage you to join the **ATX—Astronaut Training Experience.** Due to their interactive nature, **ATX** crews are small and advance reservations are required. Call 321-449-4400 to make your reservation. Your training day will begin at 10:00 a.m. and conclude at 4:30 p.m. The $225 per person fee includes **ATX** gear and lunch.

Dates

Open daily
Closed Christmas Day and certain launch days

Hours

9:00 a.m.–5:30 p.m.

Admission Cost

Adult	$29
Child (ages 3–11)	$19

There are many other tours that you can take while you are there. Many are included in the admission price. Some require an additional charge. Go to the Website and study the options.

Contact Information

Kennedy Space Center
Visitors Center
Titusville, FL 32899
(321) 449-4444
http://www.kennedyspacecenter.com
Space Shuttle launch schedule: http://spaceflight.nasa.gov/shuttle/future/index.html

G/A Airport Serving the Event

Titusville, FL (Space Coast Regional—TIX)

Phone: (321) 267-8780

Tower: 118.9

Runways:

18–36	7320′ × 150′
9–27	5000′ × 100′

The view when you're landing and taking off is amazing. Typically you'll take off on runway 9 which will point you straight toward the Saturn Launch Gantry, which you will see very clearly just on the other side of the Banana River. You will not reach 500 feet to make your turnout until you are halfway across the river. The phrase *getting there is half the fun* really applies to this trip!

FBO Name/Phone Number

TICO Executive Aviation

(321) 267-8355

www.ticoaviation.com

Ground Transportation

Rental car

Best Place to Stay $$$$$

Holiday Inn—Kennedy Space Center

4951 S. Washington Ave. (US 1)

Titusville, FL 32780

(321) 269-2121

(321) 267-4739 (fax)

HIKSC@bellsouth.net

Best Place to Stay $$

Days Inn

3755 Cheney Highway (Highway 50)

Titusville, FL 32780

(321) 269-4480

Best Aviation Attraction

#53

La Casa del Zorro

La Casa del Zorro is San Diego County's only Mobil four-diamond desert resort. It is a good enough reason for any traveler to come this way. For the aviator there is a very special treat in store. First let's look at where you'll be staying and then let's discuss the reason for making the trip.

Borrego Valley Airport (L08) is in the desert south of Palm Springs. During April the beauty of the desert wildflowers is well-known. You'll be surprised how nice this new blacktop is. Get there before 4:00 p.m. if you want fuel. The hotel will gladly instantly dispatch a limo to pick you up!

You'll want to come here only during the winter season, October 31 through May 15. The rest of the year it is very hot. It is the desert after all. You will be expected to wear a jacket at dinner. This is a dressy environment. The wine list is really quite something, and the food is wonderful. Come here expecting to be pampered and leave feeling refreshed. It is a good place for a romantic weekend getaway. Everything you expect to find at an upscale desert hotel is here. There is no golf course.

Now this is why an aviator comes here. The California State Parks department developed an aerial tour in booklet form to give pilots and passengers a chance to enjoy the natural and historic features of Anza-Borrego Desert State Park, some of which can only be seen from the air.

The booklet is in two (separate) parts: a navigation guide, which is used by the pilot, and a scenic explanation, which may be used by the passengers. It was designed for one passenger to read aloud to the others.

The sky tour begins at the Borrego Valley Airport, traverses a 150-mile (or so) roundabout tour of the desert, and returns to the airport.

The navigation section is intended for VFR use, with headings, ground reference points, and optional GPS coordinates. The description section contains a little geography, biology, and history of the local area. The hand-drawn illustrations are adequate for reference and do not detract from the vistas to be seen out the window.

The *Sky Trail* is well thought out, informative, and entertaining—an excellent way to introduce people to flying, and to the desert. Not a bad way to spend an afternoon, either.

The booklet is economically priced at $2.50, and should be available at the airport, but rather than take the chance that they are out, you may order one from:

Anza-Borrego Sky Trail
California State Parks

Colorado Desert District Office
200 W. Palm Canyon Dr.
Borrego Springs, CA 92004
(619) 767-5311

Contact Information

La Casa del Zorro
3845 Yaqui Pass Rd.
Borrego Springs, CA 92004
(760) 767-5323
reservations@lacasadelzorro.com
www.lacasadelzorro.com/

G/A Airport Serving the Event

Borrego Springs, CA (Borrego Valley—L08)
Phone: (760) 767-7415
CTAF-Unicom: 122.8
AWOS: (760) 767-3308
Runway: 7–25 5000′ × 75′

FBO Name/Phone Number

Borrego Valley
(760) 767-7415

Ground Transportation

None required

Best Aviation Attraction

#54

The Lone Eagle Journey

This is a journey that will take you 3000 miles across America and then another 3000 miles to Paris. The adventure is to follow the route of Charles Lindbergh on the most famous airplane trip ever taken by any man—the Lone Eagle's flight to Paris. You will begin where he began in San Diego. Be certain to tuck a copy of his book, *The Spirit of St. Louis,* under your arm for comfort and guidance. Whether you fly your own plane or come via commercial carrier you will land at San Diego's Lindbergh Field. It is aptly named for this is where Col. Lindbergh began his flight to Paris, right here, at this very airport. He arrived and met Claude Ryan and Frank Mahoney at the Ryan Airlines Company. Mahoney owned the place and Ryan managed it. It was their principal designer Donald Hall who would turn Lindbergh's thoughts and Ryan's standard design into a ship that could jump across the Atlantic. Together they would build the second most famous aircraft in aviation history to Lindbergh's specification and with his help. The actual building where it was assembled was an old fish cannery. It has long since been destroyed. But it once stood on Lindbergh Field near today's Coast Guard Station. Once completed, the plane made its first test flight from Dutch Flats. The San Diego Historical Society placed a plaque at that site in 1998. It was from here that Charles Lindbergh took off headed for New York's Roosevelt Field. He set a transcontinental record on that flight.

Go to the **San Diego Aviation Museum** in Balboa Park. There you will see a replica of *The Spirit of St. Louis,* which was built by the same men that built the original. It is a must see! Next you're off to New York City. After landing you'll travel to Long Island. You're going to Garden City, New York.

Take the Meadowbrook Parkway and get off at the exit for the Roosevelt Field Shopping Center. That's right—the airport has disappeared. It was razed to make way for this shopping center in 1956. Continue down Old Country Road until you come to Bob's Store. This is where Lindbergh's takeoff roll began. He proceeded through Marshall's and finally lifted off near Fortunoff's parking garage. A large sculptured plaque is there now to commemorate what is believed to be the exact spot where his wheels left the ground. They would not touch down again until he reached Paris. It would take him 33 hours to transverse the 3600-mile distance.

You must now go to Kennedy Airport and grab your jet for Paris. You will land at Charles de Gaulle Airport. It is not your final destination. You are going to Le Bourget. Today it is the site of the Paris Air Show. A good time to make this journey will be near June 19–20, 2005. That is the next date of the Paris Air Show.

Your final destination is the **Musée de l'Air et de l'Espace,** Aéroport du Bourget BP 173, 93352 Le Bourget Cedex France. It is on the site of the Le Bourget airport which is still very much in use. There is a large exhibit dedicated to the flight of Charles Lindbergh in the museum. English-speaking docents are available to explain this exhibit and to point out the exact places were Lindbergh landed and the site of the hangar where his airplane was kept.

This trip can be very expensive or it can be affordable. The cost will depend largely on your preferences and your planning and preparation. What is important is the outcome. You will have traveled the route of the most famous aviator in history. The Wrights are celebrated as a team. It is Charles Lindbergh that we herald alone. It is Charles Lindbergh who is the Lone Eagle. For the past few days you have flown along with him and become an eagle. As we all know, eagles flock with eagles, not crows.

You accepted the challenge. You studied and planned and executed. You achieved an understanding of one of America's greatest sons.

Congratulations in advance! If you can dream it, you can do it. He did and so can you.

Should you make this trip please write me and let me know about your experience. My e-mail address is jpurner@flyingsbest.com.

Best Aviation Attraction

#55

Lone Star Flight Museum and Texas Aviation Hall of Fame

(Photo by Billy Crump.)

The **Texas Aviation Hall of Fame** honors the pioneers and heroes who have made lasting contributions to aviation in Texas. The **Lone Star Flight Museum** is its flying wing. They are housed in a 100,000-square-foot, air-conditioned facility in Galveston.

The **Texas Aviation Hall of Fame** contains an award-winning collection of over 30 historically significant aircraft from the era just before, during, and after World War II. Almost all of these aircraft are flight worthy. The collection includes a B-17 Flying Fortress, B-25 Mitchell, P-47 Thunderbolt, P-38 Lightning, F4U Corsair, SBD Dauntless, and a British MkXVI Spitfire. Since the aircraft of the **Lone Star Flight Museum** are maintained to flying condition, they are in constant demand on the air show circuit. **Lone Star** aircraft present America's rich aviation heritage to all generations at over 30 aviation events across the country each year.

This outreach mission is the defining difference between the **Lone Star Flight Museum** and most other aviation museums. If it is important that the world see, celebrate, and learn from the history of aviation, then it is imperative that the planes actually fly, that they be seen doing what they were designed to do. To hear the roar of a Corsair's engine, to see it do a low pass followed by a four-point roll on the climbout is to feel, not to imagine but to feel, what the Pacific theater of World War II was like. This is a precision tool of war from our history. Our very freedom was defended by this machine and the brave men that flew her. It is not an imagined romantic notion supported by wires as it hangs unapproachably from the rafters of a sealed building.

These aircraft are first lovingly and expensively maintained. Then they are proudly flown to air shows across these United States so that those who cannot come to Galveston, Texas, might see their history and stand a little taller as these daughters of yesterday roar by.

Twice a year, all the aircraft that are on flying status are rolled out of their Galveston hangar and flown before an assembled crowd of slack-jawed fans. All of them! Can you imagine? You don't have to, you can go, you can see them, you can hear them, you can smell them. They are your history and mine. They are worth the time and the trouble it takes to get there. This is the only museum in the United States that actually flies its collection.

Dates

Open every day except Christmas Day

Hours

9:00 a.m. to 5:00 p.m.

Admission Cost

Adults	$6.95
Seniors (65+)	$5.00
Students (5–17)	$5.00
Children under 5	Free

Discounts for groups over 20

Website

www.lsfm.org

Telephone Number

(409) 740-7722

E-mail Address

Curator@lsfm.org

G/A Airport Serving the Event

Galveston, Texas (Scholes International at Galveston—GLS)

You can land at the host airport of the **Lone Star Flight Museum.** While that's good, it gets even better. With prior permission, you can taxi to and park on the museum's ramp. At GLS there is never a charge for aircraft or vehicular parking.

You'll want to visit the Moody Gardens Complex while you're here. If you visit their Website (http://www.moodygardens.com), you'll want to see it. Be sure to block in at least a day. It is impressive and there is much to do and see.

FBO Name/Phone Number

Evergreen Aviation
(409) 741-7739

Ground Transportation

Galveston Rental Car (409) 740-4287
Enterprise Rent a Car (409) 740-0700
Hertz (409) 741-3595

Best Place to Stay $$$$$

San Luis Resort & Spa
(800) 445-0090
www.sanluisresort.com

The San Luis Resort is a recipient of the AAA Four-Diamond Award, Successful Meeting's Pinnacle Award, and Meeting News Planners Choice Award. It features five restaurants including the award-winning Steakhouse, a tropical heated pool with a swim-up bar, three all-weather lighted tennis courts, and a 40,000-square-foot Executive Conference Center. It has a lavish full-service health, fitness, and beauty facility designed to "address the needs of body, spirit and mind." There is a designer boutique and Treasure Island Kid's Club for children ages 4–12.

Best Place to Stay $

Super 8
(888) 939-8680

This luxurious motel, which is only 4 years old, has repeatedly won "A" rating for being one of the top 100 Super 8s in the United States. It is located one block from the seawall on one of the major streets in Galveston, 5 miles from the cruise terminal, and 3 miles from Moody Gardens. Free parking is available for cruise guests. Microwave and refrigerator are available in each room.

#56

March Field Air Museum

On December 19, 1979, Lt. General James P. Mullin, 15th Air Force Commander, dedicated the **March Air Force Base Museum** in California. The new museum, which was housed at the March Air Force 1930-vintage base, was filled with photographs depicting the history of the base from its founding in 1918. An aircraft park, featuring aircraft that once flew from March, was established near the museum. Through the hard work of the March Field Museum Foundation, its staff and volunteers, the **March Field Air Museum** has grown to overflow its current museum building.

A significant part of the aircraft collection is on loan from the **USAF Museum at Wright Patterson AFB.** The desert climate permits many of the planes to be displayed outside. The collection is vast and eclectic. You'll see a replica of a Nieuport 11 from the World War I era, a real P-59, which was the first operational American jet fighter, and a B-47E, made famous by Jimmy Stewart in his role in *The Strategic Air Command.*

In 1996 the **P-38 National Association** and the **475th Fighter Group** added museum buildings to the grounds. On June 15, 2000, a dedication ceremony for the new Dick Van Rennes Restoration Hangar was held. The restoration hangar is located right next to the P-38 buildings.

This is a growing Air Force museum in southern California, a beautiful part of America, having bragging rights as the incubator of the jet age.

Dates

Open daily
Closed Thanksgiving, Christmas, New Year's, and Easter

Hours

Daily 9:00 a.m. to 4:00 p.m.

Admission Cost

Adults	$5.00
Children (5–11)	$3.00
Children under 5	Free

Contact Information

March Field Air Museum
22550 Van Buren
March ARB, CA 92518
(909) 697-6602
(909) 697-6605 (fax)
http://www.pe.net/~~marfldmu/
info@marchfield.org

G/A Airport Serving the Event

Riverside, CA (Flabob—RIR)
Phone: (909) 683-2309
CTAF-Unicom: 122.8
Runway: 6–24 3200′ × 50′

This airport will remind you what "friendly" means. It has been the go to spot for southern California pilots for decades. Don't miss it!

FBO Name/Phone Number

Flabob Airport, Inc.
(909) 683-2309

Ground Transportation

Rental car

Best Place to Stay $$$$$

Mission Inn
3549 Mission Inn Avenue
Riverside, CA 92501
(800) 843-7755
info@missioninn.com
http://www.missioninn.com/

The Mission Inn has been a landmark in the California Inland Empire since 1902. This National Historic Landmark hotel was recently refurbished to its original glory and is definitely *the* place you deserve to stay. It deserves each of its AAA awarded four diamonds. Whether you decide to stay here or not, do plan to take a look at its famous Aviator's Wall.

#57

MCAS Miramar Air Show

The best military and civilian pilots soar through San Diego's skies annually at the world famous **MCAS Miramar Air Show.** The Navy's own Blue Angels will be featured flying their remarkable F-18 Hornetts. The Navy's F-14 Tomcat and the Marine Corps AV-8B Harrier will make a showing. This 3-day event has won numerous awards including the International Council of Air Shows (ICAS) award for the best Military Air Show, and it is listed in the *2000 Guinness Book of World Records* for having the Longest Wall of Fire at 2500 feet. The Twilight Air Show Great Wall of Fire is featured during the Twilight Show.

Numerous types of aircraft, from old biplanes to the newest technology of stealth fighter jets, along with the Marine Air and Ground Task Force (MAGTF) simulated combat demonstration thrill the hundreds of thousands of spectators that come each year. Over 100 hands-on static display exhibits allow the general public to see the civilian and military aviation technology up close. The air show provides a unique *Twilight Show* featuring a variety of military and civilian performers, the Great Wall of Fire, and a fireworks grand finale on Saturday night. The MCAS Miramar Air Show is free to the general public.

Its setting is key. Not that many years ago, this wasn't Marine Corps Air Station Miramar, instead it was Naval Air Station Miramar, home of "Fightertown USA." It was during the Vietnam War that Miramar met its greatest challenge—to train fighter air crews in air combat maneuvering and fleet air defense. This mission was accomplished through the creation of Top Gun, a graduate-level training school for fighter air crews. The school garnered fame throughout the military for its success. The movie *Top Gun* starring Tom Cruise, with portions of the movie filmed aboard the air station, brought worldwide fame to Fightertown USA. Look around and you'll recognize parts of this base from that movie. In 1993, a Base Realignment and Closure Committee decision recommended that Naval Air Station Miramar be redesignated as a Marine Corps Air Station. The realignment involved relocating all of the Navy's F-14 Tomcat and E-2 Hawkeye squadrons. Top Gun and the last F-14 squadron left the air station in 1996 to make way for Marines from MCAS El Toro and Marine Corps Air Facility Tustin.

Dates

Mid-October

Hours

Gates open 8:00 a.m.
Showtime 9:30 a.m.–4:00 p.m.
Twilight Air Show: Saturday 5:30 p.m.–8:30 p.m.

Admission Cost

Free admission, parking, and blanket seating is available for all shows.
Box seating $10.00
Preferred box seating $15.00

Website

http://www.miramarairshow.com/

Contact Information

(858) 577-1000

G/A Airport Serving the Event

San Diego, CA (Montgomery Field—MYF)
Phone: (858) 573-1440
Tower: 119.2
ASOS: (858) 567-4337
Runways:

10L–28R	4500′ × 150′
10R–28L	3399′ × 60′
5–23	3402′ × 150′

FBO Name/Phone Number

Gibbs Flying Service, Inc.
(858) 277-0310
122.85

Ground Transportation

Enterprise Rent A Car (on the field)
(800) Rent A Car

Best Place to Stay $$$$$

La Costa Resort and Spa
Costa Del Mar Road
Carlsbad, CA 92009
(800) 854.5000
http://www.lacosta.com/

La Costa Resort and Spa is a legendary resort known for its superb guest rooms and meeting spaces, nestled amidst the beautiful coastal climate of Carlsbad, California. With 473 guest rooms to choose from, all decorated in beautiful Spanish-style theme, guests will lose themselves in comfort and style. It boasts the finest in golf and tennis facilities, with two championship golf courses and 21 tennis courts on more than 400 lush green acres. It's a mecca of fun and relaxation combined. The full-service Spa and Salon make for an even more enjoyable experience.

Best Place to Stay $

Best Western Inn, Miramar/San Diego
9310 Kearny Mesa Road
San Diego, CA 92126
(858) 578-6600
http://www.bestwestern.com/

Best Aviation Attraction

#58

MCAS Yuma Air Show

In late February each year more than 40,000 people will show up at MCAS Yuma. Built in 1928, it didn't officially become **MCAS Yuma** until July 20, 1962. Today it is the busiest air station in the Marine Corps. Fans come aboard to see one of the best air shows in this part of the country.

This annual event offers visitors the opportunity to learn and experience the objectives of the Marine Corps' Air Wing, view military and vintage planes, and other military hardware. This winning attraction promotes pride in our military and the American way of life. Occurring in February it is always one of the first air shows of the calendar year. The presentation typically includes:

- Aircraft static displays
- Marine Air/Ground Task Force demonstration
- Army Golden Knights Parachute team
- Army Yuma Proving Ground displays
- Air Force fighter demonstration
- U.S. Military fighter jets
- Attack helicopters
- Navy SEALS
- VSTOL attack jet (AV-8B Harrier)
- Military Combat Ground Unit displays

An opportunity to come aboard a Marine Corps base is always welcome. It is spit and polish to the max. You'll be interested to learn that a runway has been lengthened to 13,300 feet, making it the longest runway in the state of Arizona. The concrete used to construct it is enough for 37 miles of highway. A typical private aircraft can land and take off in its length about five times!

Dates

Annually on a late February weekend

Hours

Gate opens 8:00 a.m.
Event ends 5:00 p.m.

Admission Cost

Free

Contact Information

MCAS Yuma Air Show
Marine Corps Air Station
Yuma, AZ
(928) 269-3245
(928) 344-5592 (fax)
airshow@usmc-mccs.org
www.yuma.usmc-mccs.org/yumaairshow.htm

G/A Airport Serving the Event

Yuma, CA (MCAS/Yuma International—YUM)
Phone: (520) 726-5882
CTAF-Unicom: 119.3
Runways:

3L–21R	13,300′ × 200′
3R–21L	9241′ × 150′

FBO Name/Phone Number

Sun Western Flyers
(520) 726-4715

Ground Transportation

Rental car

Best Place to Stay $$

Days Inn
1671 E. 16th Street
Yuma, AZ 85365
(928) 329-7790

Best Place to Stay $$

Holiday Inn Express
3181 S. 4th Avenue
Yuma, AZ 85364
(928) 344-1420
(928) 341-0158 (fax)
hiexyuma3181@aol.com

Best Aviation Attraction

#59

Microsoft Flight Simulator 2004

Today's world is largely driven by the personal computer. I suppose that is not a news flash. What is amazing is the many ways they have come to touch our lives. It turns out that we can do many, many things with them so much better than we can do those same things without them. Today computers affect every aspect of aviation, none more than training. The best way I know of to gain an in-depth knowledge of planes and flying them is to latch on to the latest version of Microsoft's award-winning product Flight Simulator.

Microsoft Flight Simulator 2004 is about as far from their first attempt that popped out of the cellophane over 20 years ago as a 747 is from the Wright Flyer. It is a wondrous technological achievement on many levels. Forget the technology and focus on the training. With it you can study the history of aviation by reading the articles written by one of aviation's best writers, Lane Wallace. She'll tell you about her firsthand experience with some of aviation's greatest aircraft. You'll be able to fly each of them as a video game, if you wish, or by using **Microsoft Flight Simulator 2004** to its fullest as they are actually flown. You see Microsoft teamed up with aviation's best flight training duo, John and Martha King, to teach you how to use **Microsoft Flight Simulator 2004.**

Microsoft added renowned aviation author, lecturer, humorist, and flight instructor Rod Machado to teach you how to fly. Trust me, if you take the time to take these computer-driven classes, you will be very close to being able to fly a real airplane, not just the one inside your computer. Rod fully and artfully progresses you through the ranks of student, private, instrument, commercial, and air transport courses. Everyone can benefit from this training, particularly licensed pilots.

When you've finished Rod's flying lessons, it's time to take to the air. **Microsoft Flight Simulator 2004** will have you believing that you really are flying. The real-time weather, scenery, and audio interaction with air traffic control comes together to put you in the air without leaving your den. You'll be able to pick from a wide variety of aircraft; Wright Flyer and 747 included. My favorite is the Cessna 182. It is the closest simulation they have to the aircraft I actually own and the panel is an exact match. I use it to stay on my toes with IFR flying. The cool thing about **Microsoft Flight Simulator 2004** is that you have the ability to do flight planning just as it is actually done, then hit the go button. If you want to take off from Orlando Executive Airport, fly to Lakeland Linder Field, and shoot the ILS approach to runway 5, you can do it, using the approach plates you own or the ones that are supplied by **Microsoft Flight Simulator 2004.** Then reverse the course and return to your home airport.

If you're really into history and have tons of time on your hands you can fly the actual route Lindbergh flew in his ship *The Spirit of St. Louis.* If you get tired along the way you can do something he wished to do, pause the flight, grab 8 hours of good sleep and fire it up exactly where you left off.

To get the most out of **Microsoft Flight Simulator 2004,** I recommend that you purchase yoke and rudder pedals to make the interaction with the simulator real-world. Some folks use a joy stick and fly this product fighter-pilot style.

If you are a pilot, buy this product to stay current and begin studying with aviation's best instructor for your next rating. If you want to be a pilot or just want to learn more about airplanes this is the best investment you will ever make.

Dates

Every day

Hours

From sun up 'til the wee hours

Admission Cost

FltSim 2004	$55.00 MSRP (buy it from Amazon at a discount!)
Yoke and pedals	about $250.00
Joy stick	$20.00 to $125.00

Contact Information

http://www.microsoft.com/games/flightsimulator/

G/A Airport Serving the Event

The virtual one in your den!

Ground Transportation

None needed!

Best Place to Stay $$$$$

At home or in the office!

Best Aviation Attraction

#60

Mid Atlantic Air Museum

The **Mid Atlantic Air Museum** is located on Carl A. Spaatz Field in Reading, Pennsylvania. Open 7 days a week, it offers an interesting collection of award-winning warbirds, classic airliners, rare and unusual military and civilian aircraft, and historic exhibits. The good and bad news is that the museum's aircraft are regularly seen on the air show circuit. Call ahead to make certain the aircraft you are most interested in seeing will be in town.

My advice is that you skip a trip to the museum and come instead for the **Mid Atlantic Air Museum's World War II Weekend Air Show.** It is fabulous and should not be missed by anyone living in this part of the country. Would I fly in from Florida for it? Yes, I would. This show gives you a real flavor for the life of a World War II air warrior and the folks he left behind.

It is an annual event and is next scheduled for June 4–6, 2004. The tag line from the Museum's advertising is "Spend a Weekend in 1944." It is an honest appraisal. Over the past 15 years, it has become the largest, best-attended and most unique air show in the nation. What began as the re-creation of a World War II airfield has grown to encampments representing the European and Pacific theatres of war, an extensive homefront display, more than 70 period aircraft, 200 vehicles, and over a thousand living history reenactors and entertainers.

This is one of only five *must see* air shows. Of the ones that focus on the warbirds of World War II this is the *only* one for you. If you are looking for a *great* family getaway weekend, this is it. What if you live in Los Angeles? Well, there are great deals on airfares from LAX to New York, latch on to one and get here. You'll thank me.

Dates

Open daily
Closed major holidays

Hours

9:30 a.m.–4:00 p.m.

Admission Cost

Museum

Adult $6
Child $3

World War II Weekend

Adult (1-day)	$15
Adult (3-day pass)	$45
Child (1-day)	$5

Contact Information

Mid Atlantic Air Museum
11 Museum Drive
Reading, PA 19605
(610) 372-7333
fpierce@maam.org
http://www.maam.org/

G/A Airport Serving the Event

Reading, PA (Reading Regional Airport/Carl A. Spaatz Field—KRDG)
Phone: (610) 372-4666
Tower: 119.9
ASOS: (610) 372-9863
Runways:

18–36	5151′ × 150′
13–31	6350′ × 150′

Request a taxi to the museum (located on the north ramp of the airport).

World War II Weekend Airshow 2004
Check NOTAMS for the field!
Friday no waiver. Expect traffic but fly in and out anytime.
Saturday and Sunday
Airport closes noon to 5 p.m.
Arrive early (0730–0930) to avoid congestion.
Bring your own tie-downs. You will be directed to the transient parking area. After shutting down on a hard ramp your aircraft will be parked on grass.
With current security concerns plans could change.

FBO Name/Phone Number

Aerodynamics of Reading, Inc.
(610) 373-3000

Ground Transportation

Rental car

Best Place to Stay $$$

The Inn at Reading
1040 Park Road
Wyomissing PA 19610
(610) 372-7811

Best Place to Stay $$$

The Lincoln Plaza
5th and Washington Streets
Reading, PA
(610) 372-7777

#61

Mid-America Air Museum

Liberal, Kansas' **Mid-America Air Museum** is America's fifth largest air museum. The museum presents, preserves, and interprets the history of flight through its collection of more than 101 aircraft and related artifacts. It strives to continue to be recognized as an outstanding destination for visitors. The museum is proud to be one of Kansas' most-visited tourist attractions, with visitors from all 50 states and Germany, South Africa, Ireland, England, Mexico, New Zealand, Canada, Australia, and many other countries.

The idea for the **Mid-America Air Museum** was formed in 1986 during the 40th anniversary reunion of the **Liberal Army Airfield Veterans.** This was no idle talk for the very next year the museum was established in its current location, an old Beech aircraft facility. Retired U.S. Air Force Colonel Tom Thomas, Jr., helped things along with his 1997 donation of 56 planes. Later that same year, the City of Liberal took over operation of the facility. The collection includes fixed-wing military aircraft and helicopters from World War II, Korean Conflict, Vietnam War. I am most impressed with its fine display of production and homebuilt civilian airplanes.

If you are fortunate enough to live close by and have children, you'll want to enroll them in Camp Falcon. This is a 7-week summer educational program for elementary school and junior high kids. Campers learn all about flight throughout history, from kites and balloons to space shuttles. They perform experiments, make models, and conduct flight tests. The cost is an astoundingly low $7 per week or $49 for the entire camp!

Mid-America Air Museum is a terrific Great Plains weekend getaway. If you're a pilot, the good news is that the museum is located right on an active general aviation airport. This makes it a must see for any pilot transversing America!

Dates

Open daily
Closed Thanksgiving, Christmas, and New Year's Day.

Hours

Monday–Friday	8:00 a.m.–5:00 p.m.
Saturday	10:00 a.m.–5:00 p.m.
Sunday	1:00 p.m.–5:00 p.m.

Admission Cost

Adults	$5.00
Seniors (62 and older)	$4.00
Students (6–18)	$2.00
Children	Free

Contact Information

Mid-America Air Museum
2000 W. Second Street
Liberal, KS 67905-2199
(620)624-5263
(620) 624-5454 (fax)
liberalcityam@swko.net
http://www.liberalairmuseum.com/

G/A Airport Serving the Museum

Liberal, KS (Liberal Municipal Airport—LBL)
CTAF-Unicom: 122.8
AWOS: (620) 624-1221
Runways:

17–35	7101′ × 150′
3–21	5722′ × 150′

The museum is located on this airport. If you're flying in, be sure to phone ahead for permission to park on the museum's ramp.

FBO Name/Phone Number

Lyddon Aero Center
(620) 624-1646
(800) 659-1646
(620) 624-0566
lyddonac@swko.net

Ground Transportation

Rental car

Best Place to Stay $$:

Holiday Inn Express
1550 N. Lincoln Avenue
Liberal, KS 67901
(620) 6249700

#62

Mountain Flying

Nature wrinkled some of our land beautifully. It is an ideal place to flight see, camp, fish, hunt, and hike. Getting into the back country is best done in a small airplane, because the distances are great and roads are nonexistent. Not only is the land wrinkled, it is tall. Mountains reach above 14,000 feet, and landing strips are commonly above 7000 feet and are surrounded by jutting canyon walls waiting to clip the wing of an untrained pilot and bring him and his crippled craft back to earth in a hurry. Turbulence is born here. The bumps and jolts are rivet-rattling experiences that can upset a small airplane. Canyons are narrow and cliffs are high. Backcountry air strips are short and rough and can dogleg left or right or both just like a golf course. All things considered this is not the kinda' place a flat-land pilot from Florida is prepared to handle. But I could be, and if I was, I could conquer the back country and partake of all it provides. I want to fly here and go exploring, but I am not trained to; how about you?

Fortunately, there are those who know this country and these skies well, and they make a business of training people like us how to handle it all. Before you tackle the big sky country, stop here and take one of their seminars. It will make you and your passengers safe and comfortable. Knowing what to do is the key to survival in this environment. For example, there are three choices as to where you fly in a canyon, and two of them can hurt you. Strong updrafts and downdrafts are companions of every takeoff and landing. Learning how to use them and not fear them is essential.

Idaho is the place to go. McCall, Idaho, is home to McCall Mountain Canyon Flying. It is considered by many to be the Harvard of the bumps!

Here's their **2004 Seminar Schedule:**

June 21–25	Advanced Course—Sulphur Creek
June 29–July 2	Beginning Course
July 6–9	Beginning Course
July 13–16	Beginning Course
July 19–23	Advanced Course—Sulphur Creek

They are based at McCall Airport (MYL) in McCall, Idaho. It is the ideal launching point for excursions in the Idaho back country, with Hell's Canyon, Sawtooth Recreational Area, and the Frank Church River of No Return Wilderness Area all within easy reach.

Students train in their own aircraft which is good. Your ship must be high performance, meaning more than 200 horsepower. Exceptions can be made on an individual basis, such as Super Cubs, Scouts, Huskys, Hawk XP's, and some Maules. Owner/pilots with aircraft of less than 200 horsepower can rent one of the school's airplanes for the training or inquire about special one-on-one training in their aircraft.

Admission Cost

Beginning Seminar	All inclusive—$1495/per student
Advanced Seminar (Sulphur Creek)	All Inclusive—$2280/per student

Lodging is provided in newly remodeled cabins, with meals, guest ranch and guide services, horseback riding, live entertainment, fishing, and fun! Nonstudent guest rates are available for the duration of the course—$1055/guest.

Contact Information

McCall Mountain Canyon Flying, LLC
P.O. Box 1175
McCall, ID 83638
(208) 634-1344
admin@mountaincanyonflying.com
http://www.mountaincanyonflying.com/

G/A Airport Serving the Event

McCall, ID (McCall Airport—MYL)
Phone: (208) 634-1488
CTAF-Unicom: 122.8
AWOS: (208) 634 7198
Runway: 16–34 6107′ × 75′

FBO Name/Phone Number

McCall Air Taxi
(208) 634-7137

Best Aviation Attraction

#63

Museum of Aviation at Robins AFB

(Courtesy The Museum of Aviation at Robins AFB.)

The **Museum of Aviation at Robins AFB** is referred to as *Crown Jewel Georgia.* It is the second largest museum in the U.S. Air Force and the fourth largest aviation museum in the country. It is made unique by many things. I believe one of the most important is its exhibit of the Second World War's 14th Air Force, also known as the *Flying Tigers.* The 14th Air Force exhibit comes alive through the artifacts, memorabilia, and photographs of the thousands of men and women who served in the China-Burma-India theater during World War II. The American Volunteer Group was disbanded and replaced by the China Air Task Force whose fighter and bomber squadrons successfully defended the bases and eastern end of the India-China air supply route over the Himalayas. The 14th Air Force replaced the China Air Task Force commanded by Major General Chennault on March 10, 1943. The Flying Tigers of the small 14th Air Force compiled a record of aerial victories of more than 10 to 1 and played a significant role in the defeat of Japan.

Their comrades in the China-Burma-India Theater, *The Hump Pilots,* are also recognized here. It was their job to ferry tons of supplies over the tallest and most expansive mountain range on earth, the Himalayas. They did it successfully in everything from converted B-24s to C-47s. Hard enough to fly this route in unpressurized flying machine, but to be swarmed on by Zero's while doing it, shows unimaginable bravery. Amazingly enough this is the only place you can come to learn of it.

The museum attracts over 700,000 visitors a year. There are hundreds of exhibits and 93 remarkably restored aircraft on this beautiful 43-acre site. The museum has become a significant education and cultural center.

Along with housing the Museum of Aviation's award-winning education programs, this 65,000-square-foot hangar contains an interactive theater and exhibit hall, as well as several restored aircraft. It is also the home of the Georgia Aviation Hall of Fame.

Dates

Open every day except Thanksgiving, Christmas, and New Year's Day

Hours

9:00 a.m.–5:00 p.m.

Admission Cost

Free

Contact Information

Museum of Aviation Foundation, Inc.
P.O. Box 2469
Warner Robins, GA 31099
Phone: (478) 923-6600
Fax: (478) 923-8807
Patrick M. Bartness, President and Chief Operating Officer
pbartness@museumofaviation.org
http://www.museumofaviation.org/

G/A Airport Serving the Event

Macon, GA (Middle Georgia Regional Airport—MCN)
Phone: (478) 788-3760
Tower: 128.2
AWOS: (478) 874-8825
Runways:

13–31	5001′ × 150′ concrete
5–23	6501′ × 150′ grooved concrete

Being able to land at Robins AFB would be better, but it is not possible to get landing permission. Fortunately, MCN is only 5 minutes away.

FBO Name/Phone Number

Lowe Aviation
(478) 788-3491
www.loweaviation.com

Ground Transportation

Enterprise Rent A Car
(800) 736-8222

Best Place to Stay $$$$$

Residence Inn by Marriott
3900 Sheraton Drive
Macon, GA 31210
www.mariott.com

Best Place to Stay $

Holiday Inn—Warner Robins
2024 Watson Blvd.
Warner Robins, GA 31093
www.holiday-inn.com

Best Aviation Attraction

#64

Museum of Flight

The rounded shape of the Museum's mock air traffic control tower contrasts with the distinctive needle nose of Concorde. (*Photo by William Anthony, courtesy of the Museum of Flight.*)

One thing is for certain, this museum is well-funded: It sits on a 12-acre site on Seattle's Boeing Field/King County International Airport, which is shared with several aviation-oriented companies. The largest is the Boeing Aircraft Company. Perhaps their presence explains the 10,000-foot-long by 200-foot-wide runway. This is a museum that a private pilot can come to easily. You are invited to land and taxi over to the museum's ramp and park. If you need fuel or other services, there are several FBOs to choose from. My personal pick is Galvin.

The museum is housed in two buildings: a large 273,000-square-foot gallery and a small 35,000-square-foot library. The glass-and-steel main building soars to six stories. You'll be struck by 22 aircraft hanging from the ceiling. One of these is a DC-3! It is the most impressive display of aircraft I have ever seen.

This is a museum that focuses on visitor involvement. You may sit in the cockpit of a real SR-71 Blackbird or F/A-18 Hornet if you prefer. America's first jet-powered Air Force One is here. The first President to use it was Dwight Eisenhower. You may board it. Recently, a Concorde has been added. It sits outside on the ramp and is available for visitor boarding, whenever it isn't raining. It rains a lot in Seattle so don't count on boarding.

You'll be interested in seeing The Red Barn. This two-story wooden structure was erected in 1909. The National Register of Historic Places credits it as the birthplace of the Boeing Airplane Co. and professes that it is the oldest existing aircraft manufacturing facility in the United States. I know that a 747 won't fit in it, but I'll bet it would easily fit inside a 747.

There is a lot of construction going on as a new wing will be opened mid-2004. World War I and II airplanes will be displayed here. Hung from the ceiling? Perhaps, but I really don't know.

The museum's Restoration Center is located at Paine Field/Snohomish County Airport in nearby Everett, Washington. The public is welcome there as well (Tuesday through Thursday from 8 a.m. to 4 p.m. and Saturday from 9 a.m. to 5 p.m.). The receptionists at the museum's main desk can provide directions.

The story of flight is told on film in the intimate William M. Allen Theater. It seats approximately 250 people and is capable of projecting any film format—16, 35, and 70 mm film as well as project data and images through state-of-the-art video projectors. A variety of film programs are offered daily. Check the schedule as the programs change frequently.

(Courtesy the Museum of Flight.)

You'll want to spend most of a day here, longer if you poke around the library doing research for the definitive American aviation novel that you plan to write one day. Quence your thirst and appease your appetite at the Wings Café. Sit near a window for a up-close view of Boeing Field. The Café is open during regular gallery hours and museum admission is not required. It is a great place for a fly-in lunch!

Best of all, when you're tired of looking at airplanes and reading about their pilots you can take a flight in an aviation ghost. **Olde Thyme Aviation** maintains and operates a fleet of open-cockpit aircraft. They are based at Galvin aviation and they will pick you up from the museum. You and a friend can suit up and go barnstorming around downtown Seattle for as little as $125. This is the perfect way to end or begin a day at any aviation museum.

My guess is that one day this museum will be the premier aviation museum in the world. All it takes is cash and dedication. They appear to have both in abundance.

Dates

Open seven days a week

Closed Thanksgiving Day and Christmas Day

Hours

Daily from 10:00 a.m. to 5:00 p.m.

The first Thursday of each month 10:00 a.m. until 9:00 p.m.

Admission Cost

Children (4 and under)	Free
Youths (5–17)	$6.50
Youth groups*	$5.50
Adults (18–64)	$11.00
Adult groups* or seniors (65+)	$10.00

*Groups: 10 or more paid visitors paying in one transaction.

Thanks to the generosity of Wells Fargo Bank, a sponsor, admission is free the first Thursday evening of every month from 5:00 p.m. until 9:00 p.m.

Contact Information

Museum of Flight
9404 East Marginal Way South
Seattle, WA 98108-4097
(206) 764-5700
www.museumofflight.org
info@museumofflight.org

G/A Airport Serving the Event

Seattle, WA (Boeing Field/King County International—BFI)
Phone: (206) 296-7380
Website: www.metrokc.gov/airport/
Tower: 120.6
ASOS: (206) 763-6904
Runways:

13L–31R	3710′ × 100′
13R–31L	10,000′ × 200′

The museum allows private aircraft to park on their ramp! You must contact the museum's Security Department in advance to make arrangements. The phone number is (206) 764-5700.

FBO Name/Phone Number

Galvin Flying Service, Inc.
(206) 763-0350
(206) 767-9333 (fax)

Ground Transportation

Enterprise Rent A Car

800 Rent A Car

Best Place to Stay $$

Red-Lion Seattle—South

11244 Pacific Highway South Seattle
Seattle, WA 98168
Phone: (206) 762-0300
Fax: (206) 762-8306
Res: (800) RED-LION
sales@redlionseattle.com

In the heart of south Seattle's business and industrial district, this three-diamond hotel has all the amenities of a true business-class hotel. It provides complimentary airport shuttle service to Boeing Field, SeaTac Airport, the Museum of Flight, and limited service to Southcenter Mall and downtown Seattle including the Amtrak train station.

Best Aviation Attraction

#65

National Model Aviation Museum

The first practical steps toward flight began with a design concept, a drawing, then a model. So it was with da Vinci and the Wright brothers. First came the thought, then the detailed engineering design, then a model to prove the concept. Nothing has changed. The engineers at Boeing, Cessna, and Lockheed do it exactly the same way.

However, models of airplanes are often more than a stepping-stone to a bigger machine, for they can be an end in themselves. As children and adults many of us built models for their own sake—to touch, to dream with, and sometimes to watch them fly. They are the stuff dreams are made of, the windows to our imagination. In this age of high-tech everything, model airplanes are very advanced. In 2003 a group of very patient and persistent men flew one all the way across the Atlantic. Following the flight of Lindbergh? No, blazing a trail of their own, with a pilotless plane flown only by the dreams of its designers and builders steering a course derived by computer and monitored by a constellation of man-made satellites. Are they models, too?

If you have ever admired models of airplanes. then a visit to the National Model Aviation Museum would be a special moment for you. For here, hundreds of models are enshrined, but not just any models. Here only the best models are kept. Every variety is represented. Static display types and powered ones ready to launch. Each was a trendsetter in its day and each is the *best* of class.

Plot your course to Muncie, Indiana, where the largest collection of model aircraft in the United States awaits. Discover how aeromodeling has changed the world of aviation. Examine the astonishing craftsmanship and artistry of its masters. During the summer months, the adjacent 1000-acre flying site comes alive with members of the **Academy of Model Aeronautics** flying their aircraft in competitions. It is best to come during the national championships which are held in July and August.

Dates

Open every day
Closed Sundays Thanksgiving through Easter

Hours

Monday through Friday 8:00 a.m.–4:30 p.m.
Saturday and Sunday 10:00 a.m.–4:00 p.m.

Admission Cost

Adults	$2
Youths (7–17 years)	$2
Children	Free

Contact Information

National Model Aviation Museum
Academy of Model Aeronautics
International Aeromodeling Center
5151 East Memorial Drive
Muncie, Indiana 47302
www.modelaircraft.org
(765) 287-1256

G/A Airport Serving the Event

Muncie, IN (Delaware County Airport-Johnson Field—MIE)
Phone: (765) 747-5690
Tower: 120.1
ASOS: (765) 288-9617
Runways:

14–32	6500′ × 150′
2–20	4998′ × 100′

FBO Name/Phone Number

Muncie Aviation Company (MIE)
(765) 289-7141

Ground Transportation

Rental car

Best Place to Stay $$

Lees Inn
3302 N. Everbrook Ln.
Muncie, IN 47304
(765) 282-7557

Best Place to Stay $

Best Western
3011 W. Bethel Avenue
Muncie, IN 47304
(765) 282-0600

Best Aviation Attraction

#66

National Museum of Naval Aviation

The Navy's frontline fighter, the F-14 Tomcat, graces the front entrance. (© *1997 Naval Aviation Museum Foundation.*)

The **National Museum of Naval Aviation** is one of the three largest aviation museums in the world. The interior measures just under 300,000 square feet. It is here that the 9-decade-long story of naval aviation is told with 150 perfectly restored aircraft. You'll see everything, beginning with the Navy's first airplane, a Curtis Triad.

Loaded with the most advanced technology, the museum is prepared to enhance each visitor's experience. Have you ever wanted to know what it felt like to fly the Navy way? The museum's motion-based flight simulator can launch you from a carrier, take you on a Desert Storm strike mission, or let you ride with Blue Angels. It is extraordinary!

In 1996, an IMAX theater was added. The 85-foot-wide screen reaches a height of almost seven stories. The museum's signature film, *The Magic of Flight,* features in-flight sequences of the Blue Angel's made more exciting by the 15,000-watt sound system. It is shown daily from 10:00 a.m. to 4:00 p.m. every hour on the hour.

The **National Museum of Naval Aviation** is located at the historic Pensacola Naval Air Station in Florida, which was designated the first naval air facility in 1914. The Pensacola Naval Air Station trains student aviators and provides survival training all the way to the coveted Wings of Gold award. You'll want to take *The Flight Line Bus Tour* to see as much of the facility as you possibly can. It takes you on a free, 20-minute, narrated tour of the approximately 40 aircraft displayed on the flight line behind the museum's restoration hangar.

What makes this museum so special is the presence of the Blue Angels. This is their home base and they can be seen practicing from an area adjacent to the museum. Signs are posted to direct visitors to viewing and parking locations. Practices are typically held on Tuesday and Wednesday mornings at 8:30 a.m. (weather permitting) from March through October. Call the Blue Angels Public Affairs Office at (850) 452-2583 for exact dates and further information.

Dates

Every day, except for Thanksgiving Day, Christmas Day, and New Year's Day

Hours

Open 9:00 a.m. to 5:00 p.m. seven days a week

Former Blue Angel A-4 Skyhawks suspended in the Blues' famous diamond formation. (© 1997 *Naval Aviation Museum Foundation.*)

Admission Cost

Free

Websites

http://www.navalair.org
http://www.navy.com/blueangels

Telephone Numbers

(850) 452-3604 or (850) 452-3606
Fax: (850) 452-3296

E-mail address

Naval.Museum@smtp.cnet.navy.mil

G/A Airport Serving the Event

Pensacola, FL (Pensacola Regional—PNS)

FBO Name

Pensacola Aviation Center

Phone Number

(850) 434-0636

Ground Transportation

Rental car

Best Place to Stay $$$$

Portofino

(888) 909-6807

Portofino is an exclusive destination resort and spa nestled between the pristine sugary beaches of the Gulf of Mexico and the tranquil waters of Santa Rosa Sound. Its unique location, at the east end of Pensacola Beach and just 15 minutes from downtown Pensacola, offers the best of both worlds—unspoiled beaches, plus galleries, museums, and shopping within easy reach.

Inside and out, Portofino is stunning from every viewpoint...an architectural work of art with a timeless Italian Mediterranean flair. The convenient underbuilding parking allows for over 60 percent of the 28-acre resort to remain an undeveloped wildlife habitat with lush tropical landscaping. The spacious two-bedroom residences offer breathtaking water views from nearly every room.

One of the main features is Spa Portofino, an exclusive Aveda Concept Spa. Portofino also features an indoor pool and five heated pools, whirlpool spas, saunas, steam rooms, a gourmet restaurant, seasonal Kids' Camp and access to Tiger Point, a 36-hole championship golf course.

Best Place to Stay $$

Clarion Suites Resort and Convention Center

(800) 874 5303

http://www.clarionsuitesresort.com/

The Clarion Suites Resort and Convention Center is Pensacola Beach's only all-suite hotel. Enjoy the warm waters of the Gulf of Mexico and Pensacola Beach's world famous sugar white sands just outside your door. After a day on the beach, you'll want to spend some time golfing, charter boat fishing, sailing, shopping, or visiting historic sites. The Clarion Suites Resort offers the perfect setting for your family vacation.

#67

National Soaring Museum

Soaring is flight in its purest form. Its story is told a few miles from Elmira, New York, atop Harris Hill by the **National Soaring Museum** (**NSM**). The 350-acre site is shared with a county park. The park's extensive facilities include a pool, youth camp, cabins, picnic areas, camping facilities, a kiddie park, three-hole golf course, driving range, a spectacular scenic overlook, and an active historic glider port operated by the Harris Hill Soaring Corporation. Come here and spend a few days camping, golfing, learning, and even riding in a sailplane should you choose.

The **Soaring Society of America** (**SSA**) was organized in 1932. The first informal meeting was held on February 20 in the McGraw-Hill building in New York City. The **SSA**'s first mission was to host an annual soaring contest. Its purpose has since mushroomed. In the 1960s the **Soaring Society of America** formed the **National Soaring Museum** and designated it as the official archives repository for **SSA** records. All pilots' contest records, badge records, and licenses are retained here.

The **National Soaring Museum** was also challenged to preserve a unique aspect of American aviation history—winged motorless flight. Today, approximately 60 ships are housed at **NSM.** They tell the tale of soaring's past, present, and future. Through the generosity of the members and friends of the **National Soaring Museum,** it has gathered the largest collection of sailplanes and gliders in the western hemisphere. Many of them are on display here. Others have been loaned or donated to museums across the country. A number of private collections donated by some of the most significant figures in soaring history—Klemperer, Barnaby, Hall, Ryan, Lincoln, Prue, and the Schweizers—are kept at **NSM** as well.

Harris Hill Soaring Center shares the hilltop and the ridge overlooking the beautiful Chemung Valley in rural upstate New York with **NSM.** Sailplane rides are offered here on weekends April through October, and on weekdays from mid-June to Labor Day. The charges are modest, $55 per ride in a Schweizer 2-33 glider or $65 per ride in a Schleicher ASK-21 high-performance sailplane. Each ride lasts about 20 minutes.

Dates

Open daily
Closed Thanksgiving Day, Christmas Eve and Christmas Day, New Year's Eve and New Year's Day

Hours

10:00 a.m. to 5:00 p.m.

Admission Cost

Adults	$6.00
Seniors	$5.00
Children	Free

Contact

National Soaring Museum
51 Soaring Hill Drive
Elmira, NY 14903-9204

G/A Airport Serving the Event

Elmira, NY (Elmira/Corning Regional Airport—ELM)
Tower: 121.1
ASOS: (607) 796-0065
Runways:

6–24	7000′ × 150′
10–28	5402′ × 150′

Ground Transportation

Rental car

Best Place to Stay $$

Holiday Inn Elmira
1 Holiday Plaza
760 E. Water Street
Elmira, NY 14901
(607) 734 4211

#68

N'Awlins Air Show

N'Awlins Air Show is located at the **Naval Air Station, Joint Reserve Base,** which is just outside Belle Chasse, Louisiana. You enter the base via the Barriere Road Gate, but don't drive to the show. The roads are narrow and congested on a good day; on show day they are impossible. You can pick up a shuttle downtown on the Canal Street side of the Quarter. Most downtown hotels will have transportation suggestions for you. Take them! Basically, this is your standard military-supported air show. This one is a double treat because of its location, New Orleans, and because the Blue Angels are going to be the 2004 headliners. They are terrific!

Friday is practice day. Plan to take advantage of it. If there is a day to drive to the base this is it. Call ahead for details. The crowds will be very, very light and you will get to see the Blues take off, practice their entire act, and land. If you are in luck and you probably will be, their support aircraft will practice their high-performance takeoff on Friday. It is a sight!

The next thing I like about this show is the hospitality of the base. It is truly southern. You can even go to their on-base Hangar Deck Café. Do it! The food is good and the ambiance is pure Navy.

Do not stay in this part of town. It is covered with oil refineries and chemical plants and there isn't anything to see. Plus you are in the tourist mecca of the South. Enjoy it! Stay in the French Quarter. Be a tourist. You are one after all. Drink very little and eat a lot. New Orleans is about food. All of my friends in New Orleans have the same outlook: life is endured between meals. Life is about eating! I swear you will never be long in the company of a real New Orleans' resident before he or she asks you where you ate last night and what you had. They care deeply about food and they should. The best food in the world is served here. Everywhere! If you get a bad meal it's your fault.

Dates

October 23–34, 2004

Hours

Gates open 7 a.m.

Admission Cost

Free

Contact Information

Naval Air Station Joint Reserve Base New Orleans
400 Russell Avenue
New Orleans, LA 70143-5012
(504) 678-3710
http://www.mwrneworleans.com/N'Awlins_Airshow.htm

G/A Airport Serving the Event

New Orleans, LA (Lakefront—NEW)
Phone: (504) 243-4010
CTAF-Unicom: 119.3
AWOS: (504) 242 5993
Runways:

18L–36R	3699′ × 75′
18R–36L	6879′ × 150′
9–27	3094′ × 75′

This is a great general aviation airport. Certainly, it benefits from being very near to downtown New Orleans and the famous French Quarter. The prevailing northerly winds will have you on one of the north-bound parallel runways for takeoff. Believe the signs at the end of the runway. You will indeed lose all horizontal visual input immediately after takeoff. The haze and the lake will see to that. Be prepared to be on the gauges.

FBO Name/Phone Number

General Aviation of N. O.
(504) 241-2700

Ground Transportation

Rental car

Best Place to Stay $$$$$

JW Marriott Hotel—New Orleans
614 Canal Street
New Orleans, LA 70130
(504) 525-6500

Best Place to Stay $$$

Renaissance Arts Hotel
700 Tchoupitoulas Street
New Orleans, LA 70130
(504) 613-2330

#69
NBAA Annual Meeting & Convention

The **National Business Aviation Association, Inc. (NBAA)** will hold its 57th annual convention in Las Vegas, Nevada in October 2004. This annual event is the business aviation industry's largest gathering of buyers and sellers. The NBAA is made up of 7500 member companies. Each member company operates or shares the operation of at least one aircraft. The public is invited to attend the exhibitors hall during the convention and get a peak at the latest business-capable aircraft and equipment being offered by the 1000 or so manufacturers who are hoping to find favor with the nearly 30,000 NBAA attendees. It is the only opportunity to do direct, side-by-side comparisons of nearly every product and service available to company flight departments.

What exactly is business aviation? It is part of general aviation, meaning everything that flies not for the airlines or the military. Business aviation specifically is any general aviation operator who uses an aircraft as a tool in the conduct of business.

This includes individuals flying single-engine, piston-powered airplanes as well as the largest multinational corporations operating a fleet of multiengine, turbine-powered aircraft. Of the 9352 aircraft operated by NBAA Members, jets weighing 29,999 pounds or less are the most popular. These 3285 light and medium jets constitute 35 percent of the NBAA fleet. The Citation and the Learjet are the most widely owned aircraft in this group.

It is interesting to note that the most numerous additions to the jet fleet recently have been on the bottom rather than the top end of the market. Typically these are six-place aircraft selling for less than $3 million. Some are even being touted at less than $1 million. This opens personal jet travel to many more individuals and companies.

Your reason for going to this convention may be simple idle curiosity to see how the dwellers of the corner office get around. Yes, you will be able to board and explore most of the exhibits and you will be given tons of literature. You might go because you are a piston-engine aircraft owner who is considering moving up and wants to see what is available. Any reason is a good reason to see more of this important segment of general aviation.

Dates

October 12–14, 2004

Hours

Opens 10:00 a.m.
Closes 4:00 p.m.

Admission Cost

Exhibits: $25.00

Contact Information

National Business Aviation Association, Inc.
1200 18th St., NW
Suite 400
Washington, DC 20036-2527
(202) 783-9000
(202) 331-8364 (fax)
info@nbaa.org
http://www.nbaa.org/

G/A Airport Serving the Event

Las Vegas, NV (North Las Vegas—VGT)
Phone: (702) 261-3806
Tower: 125.7
ASOS: (702) 648-6633
Runways:

12–30	5000′ × 75′
7–25	5000′ × 75′

FBO Name/Phone Number

North Las Vegas Airport
(702) 261-3800

Ground Transportation

Rental car

Best Place to Stay $$$$$

So, so many

Best Place to Stay $:

Even more

#70

New England Air Museum

New England Air Museum has some truly rare aircraft on display. My all-time favorite plane is the Republic RC-3 Seabee. I never got to fly one and if I could find one on the market I'd quickly offer my Cherokee in trade. They have a Bleriot X1 Monoplane here as well; this is not a replica!

The museum offers an audio tour, which is really novel for an aviation museum. I hope others are encouraged to follow New England's lead as this really makes the exhibits come alive. The tour gear consists of a digital handset and a map of the two indoor hangars. You use a supplied map to locate the numbered pylons around the museum. Next you type the pylon number into the handset and listen to the in-depth narration. This fascinating tour takes visitors through the history of flight chronologically from man's earliest recorded attempts at engineering a flying machine through the present. The tour has 30 stops and takes approximately 1 hour to complete. The $3 rental fee is well worth the price. I completed the audio tour, turned in my handset and went back through the museum to see the displays that weren't covered.

There are three best times to come to this museum each year. They are called ***Open Cockpit Days.*** You will be invited to climb into the pilot's seat of World War II fighters, jet fighters, an airliner, helicopters, and a civilian plane. Bring your camera; this is a great chance to get some photos of these old birds' interior. You may never have a chance like this again, so don't let it pass you by.

Dates

Open daily
Closed Thanksgiving, Christmas, and New Year's Day

Hours

Daily 10:00 a.m.–5:00 p.m.

Admission Cost

Adults	$8.00
Seniors (60 and up)	$7.00
Children (6–11)	$4.00
Children (5 and under)	Free

Contact Information

New England Air Museum
Bradley International Airport
Windsor Locks, CT 06096
(860) 623-3305
(860) 627-2820 (fax)
staff@neam.org
http://www.neam.org/

G/A Airport Serving the Event

Windsor Locks, CT (Bradley International—BDL)
Phone: (860) 292-2000
Tower: 120.3
ASOS: (860) 627-9732
Runways:

6–24	9510′ × 200′
15–33	6857′ × 150′
1–19	5145′ × 100′

The museum is on the field.

FBO Name/Phone Number

Signature Flight Support
(860) 623-394
Signature offers a free shuttle bus to and from the museum. Please call to make arrangements in advance.

Ground Transportation

Rental car

Best Place to Stay $$$

Sheraton Hotel—Bradley International Airport
Bradley International Airport
Windsor Locks, CT 06096
(860) 627-5311
(800) 325-3535

Best Aviation Attraction

#71

Octave Chanute Aerospace Museum

In 1917 Rantoul, Illinois, was chosen by the U.S. Army Air Service to train pilots for World War I. Over the next 75 years it became the premier technical training school for the USAF and trained over two million men and women. In 1993 Chanute Field Air Force Base (the third oldest active base in the U.S. Air Force) was closed, converted to civilian use, and renamed Frank Elliot Field. The **Octave Chanute Aerospace Museum,** the largest aerospace museum in Illinois, is located on the airport grounds of Frank Elliot Field. Its mission is to display and preserve the history of Chanute Air Force Base, the United States Air Force, and Illinois aviation.

The crown jewel of the museum's collection is the world's only remaining Boeing XB-47 Stratojet prototype. During your visit you will be thrilled to see replicas of Charles Lindbergh's *Spirit of St. Louis* and Chanute Field's first aircraft, the Curtiss Jenny biplane.

For me the visit to the **Octave Chanute Aerospace Museum** is made worthwhile by the opportunity to go inside the cockpit of a Boeing B-52D Stratofortress. Don't forget to bring your camera. Next you can enter the cargo bay of a Lockheed C-130 Hercules. Those of you who lived through the cold war will be very interested in the underground tour of a Minuteman ICBM training silo.

The **Octave Chanute Aerospace Museum** is fulfilling its educational mission by creating and operating the **Chanute Aviation Camp.** It is geared to 7th through 12th grade students and is held annually during the month of June. Classes are held in the Discovery Learning Center at the museum, in the museum's hangar, and at Willard Airport. The University of Illinois Institute of Aviation cosponsors the camp and provides great insight to the campers' curriculum.

Campers will learn about aviation, aerospace, and related careers through classroom instruction, lectures, videos, hands-on operation of equipment, and actually flying an aircraft! A tour of Willard Airport is conducted, including the FAA control tower, the firehouse, the terminal building, and aircraft maintenance facilities.

Flight instruction is given in flight simulators. Campers then take a 45-minute ride in one of the University of Illinois Institute of Aviation's aircraft.

The total cost for the 1-week, half-day-long course is $150. This is a wonderful gift for any youngster who has expressed the slightest interest in aviation.

Dates

Open daily

Hours

Monday–Saturday	10:00 a.m.–5:00 p.m.
Sunday	Noon–5:00 p.m.

Admission Cost

Adults	$7.00
Seniors	$6.00
Children (K–12)	$4.00
Children (4 and under)	Free

Contact Information

Octave Chanute Aerospace Museum
(217) 893-1613
(217) 892-5774 (fax)
admin@aeromuseum.org
http://www.aeromuseum.org/

G/A Airport Serving the Event

Rantoul, IL (Frank Elliott Field—TIP)
Phone: (217) 892-2121
CTAF-Unicom: 123.0
ASOS: (217) 892-4999
Runways:

9–27	5000′ × 75′
18–26	4895′ × 75′

Fly-ins are encouraged! The museum is adjacent to the FBO. Private and business aircraft are welcome.

FBO Name/Phone Number

Precision Aviation, Inc.
(217) 892-2121

Ground Transportation

Rental car

Best Place to Stay $$

Best Western Heritage Inn
420 S. Murray Road
Rantoul, IL
(217) 892-9292

Best Place to Stay $$

Super 8 Motel
207 S. Murray Road
Rantoul, IL
(217) 893-8888

Best Aviation Attraction

#72

Oshkosh AirVenture

(Courtesy EAA.)

The Experimental Aircraft Association (EAA) holds its annual convention at the end of July and the beginning of August each year at Wittman Regional Airport in Oshkosh, Wisconsin. They invite the public to join them. Over 700,000 accept and swell the local population 10-fold. Rightfully called *The Big Show,* it is the largest air show in the world. While it is going on, this town of 85,000 boasts the busiest airport in the world. The size of the site and the rows and rows of magnificent aircraft are simply overwhelming. Pilots from across the world flock to this gasoline-powered Capistrano, and 12,000 airplanes show up for the event. Typically about 2800 show aircraft participate at **EAA AirVenture Oshkosh,** including homebuilts, antiques, classics, warbirds, ultralights, and rotorcraft. More than 40,000 people camp at **Camp Scholler,** with an additional 5000 in transient aircraft and show-plane camping areas. More than 4500 volunteers contribute more than 250,000 hours to AirVenture each year. AirVenture attracts visitors from 80 nations and over 830 media people from 5 continents. There is something for everyone, from the tiny tot to mom.

AirVenture, or **Oshkosh,** the name by which it is best known, is about airplanes. All kinds of airplanes, military, civilian, private, commercial, all are welcome and celebrated on this flight line. You'll find the oldest of the old and the newest of the new. They all come. The roots of Oshkosh run deep into the soil of the homebuilt aircraft crowd. That is what EAAers do. They build from scratch or from a kit their own airplanes and then they fly them! Can you imagine? The point is that they can and they do. In recent years EAAers are heeding the call of not just airplanes but spacecraft as some of their members compete to be the first civilians to build craft capable of suborbital flight. Can you imagine building a spacecraft in your garage in your spare time? No! Well these folks dream that dream and then act it out. When they are still at work or near complete they bring their ideas and their machines here, to Oshkosh, to The Big Show.

If you have never been to an air show before, this is the place to start. Oshkosh is the place where dreams become reality.

Dates

July 27 through August 2, 2004

Hours

8:00 a.m. until...(a 24-hour event)

Admission Cost

All AirVenture admissions are sold at the gate.

Members*		
	Daily	Weekly
Adult	$19	$94
Spouse	$15	$47
Student (14–18)	$14	$46
Youth (8–13)	$ 9	$34
Children (7 and under)	Free	Free
Adult guest (limit 2)	$25	$168

Nonmembers		
	Daily	Weekly
Adult	$29	$203
Student (14–18)	$16	$112
Youth (8–13)	$11	$77
Children (7 and under)	Free	Free

*Annual adult membership: $40

Websites

Oshkosh Convention and Visitors Bureau: www.oshkoshcvb.org
AirVenture www.airventure.org

Contact Information

EAA Aviation Center
3000 Poberezny Road
Oshkosh, WI 54902
(920) 426-4800

G/A Airports Serving the Event

Oshkosh, WI (Wittman Regional Airport—OSH)

Wittman is a truly wonderful general aviation airport. It offers four runways, the longest of which is 8002 feet. Every IFR approach including an ILS is available. During the week of AirVenture, OSH is the busiest airport in the world; so busy, in fact, that the radios of arriving aircraft are used to receive instructions but never to transmit a request or a reply. "Yellow airplane land long, white airplane land midfield, red airplane land short," barks the controller. Somehow it works. Everyone seems to move just the right way at just the right time in this totally unrehearsed airborne ballet. Thanks to the FAA controllers who volunteer their vacations, it is safe.

If you decide to land at OSH be sure to review the approach and landing procedures. They aren't complex but they must be adhered to precisely. You'll find them on the EAA Website. They'll also be pleased to send you a paper copy if you request.

Appleton, WI (Outagamie County Regional Airport—ATW)

Outagamie County Regional Airport is just 22 miles north of OSH. Some say it is a safer choice. It is surely easier for arrivals and departures during the week of the AirVenture. With over 750 planes arriving there isn't room on the ramp for everyone; many will have to park on the grass. Place your fuel order when you arrive, you don't want to stand in line waiting for fuel when it's time to depart.

FBO Name/Phone Number

Basler Flight Service (OSH)
(800) 558-0254

Orion Flight Services, Inc. (OSH)
(866) 359-6746

Ground Transportation

When 700,000 spectators descend on Oshkosh things get busy, real busy. The hardest part of this trip could easily be getting around town. This is one of those times that one car for one person doesn't work very well. Having a car for the week of AirVenture is no great blessing. If you're staying in a neighboring village there is no way to avoid it, but if you're not leaving Oshkosh, you'll be better served by buses. They're everywhere and go everywhere, frequently and at low cost.

Rental Cars

City	Company	Phone
Oshkosh	Hertz Rent-A-Car	(800)654-3131
	Avis Rent-A-Car	(920)730-7575
Appleton	Hertz Rent-A-Car	(800)654-3131
	National Car Rental	(920)739-6421
	Enterprise Rent-A-Car	(800)RENT-A-CAR

Buses

Shuttle buses are available from the Outagamie County Airport and various motels in Appleton for less than $10. For a schedule, call Nationwide Travelers at (920) 734-5620 or visit their Website at www.nationwidetravelers.com.

The Oshkosh transit system provides shuttle buses from the downtown Transit Center and the University of Wisconsin–Oshkosh campus to the convention grounds. A separate fare of $2 (round trip) is charged for this service. For more information visit their Website at http://www.ci.oshkosh.wi.us/Transit/Transit.htm.

The Westside Shuttle will begin operating a day or two before the convention and will continue to run through the last day of the show. The first bus pickup at Camp Scholler is at 8 a.m. and last dropoff at Camp Scholler is 9 p.m. Cost is $2 for the entire day. Stops include grocery stores, shopping centers, the movie theater, laundry, restaurants/lounges, and numerous businesses along both sides of the Highway 41 frontage road.

Lodging

When Oshkosh swells from a population of 85,000 to over 700,000 for a week things get pretty interesting. Solving the housing problem strains all the local resources. Motels and hotels are booked year to year by the same guests and passed along from father to son and mother to daughter. Finding a hotel room in Oshkosh during the convention is like finding warm weather in Maine in the January. It rarely happens. Fortunately there are several good alternatives.

Beginning in March of each year you can call the Housing Hotline at (920) 235-3007 for private housing. All are within 10 miles of the convention. Local citizens rent out rooms or their entire home to EAA visitors. Rooms average between $50 and $100 per night. Homes average between $200 and $300 plus per night. Call between 8:30 a.m. and 4:00 p.m., Monday through Friday (Central Standard Time). The Housing Hotline also has information on camper and RV parking spaces.

The University of Wisconsin–Oshkosh makes as much dorm space available as it can. What a great way to relive your college days. To be placed on a waiting list call (920) 424-3226 or e-mail uwoeaa@mio.uwosh.edu. They require a three-night minimum.

Thousands of EAA members also choose to camp out. Fortunately there are wonderful facilities on and off the convention site. To stay at **Camp Scholler** you must be an EAA Member. It is located adjacent to the convention site and is for vehicle/tent camping only. Approximate campsite size is 20′ × 30′. Larger units will require multiple campsites. Shower facilities, portable toilets, RV pumping station, and portable pumping services are all available. Access to food outlets and the Country Store make camping here a breeze. Electric and water hookups are not available. The cost is $17 per night with a three-night minimum.

To use the close-in **Showplane Camping,** you must be an active EAA member, and you must have flown in on a plane eligible for EAA's Judging Program. Showplane Camping is on the flight line. One tent per aircraft is permitted and wrist bands must be worn at all times to gain access to this area.

If you don't own a showplane but want to fly in and camp out **The North 40** is the place for you. To camp in the General Aviation Camping area (North 40), you must be an active EAA member. If your aircraft is parked in this area you will be charged for camping. Be sure to bring your tie-downs as all aircraft must be tied down.

The most unique camping opportunity at the convention is offered by the **Oshkosh Community YMCA.** Located 3 miles west of EAA grounds it provides 125 *indoor* sites in their soccer arena. Pitch your free-standing tent or sleeping bags on their padded astroturf. Rates of $15–$35/night include scheduled shuttle service to EAA AirVenture. While here you'll enjoy full member privileges at a brand-new YMCA. Locker rooms, showers with towel service, café, adult pool, family pool, spa, sauna, fitness facilities, and youth programs are all included. Call (920) 230-8439 to make reservations or visit their Website, www.oshkoshymca.org.

Best Aviation Attraction

#73

Paris Air Show

The **Paris Air Show** is actually a trade show where the largest aircraft manufacturers from around the world wheel and deal with their largest customers. Major announcements are made here as new products are put through their paces above this crowd of buyers and sellers.

The event occurs every other year and 2001 was a record year for the number of exhibitors (1856), the amount of space used for the show (308,113 m^2), the number of visitors present (306,658), and the number of nations represented (42). In 2003 the numbers dropped a bit for several reasons. The war in Iraq and the poor worldwide economy loomed large in minds of many. For those that did come they noticed something very different indeed.

Throngs of Parisians crowded Le Bourget on a special evening in 1927 to get a glimpse of the first man to fly across the Atlantic. In 2003 the stars of the **Paris Air Show** (conducted at Le Bourget) were the UAVs (Unmanned Aerial Vehicles). How strange it was to see the attention shift to pilotless aircraft at the very airport where transoceanic air travel was ushered in by Charles Lindbergh. A modern version of *The Spirit of St. Louis* may soon arrive without a Lone Eagle at the controls.

Paris is a wonderful city from which to start any European vacation. If you can, time your trip for the middle of June in any odd-numbered year. Plan to stay in Paris for at least 4 days.

Just 8 kilometers north of central Paris is Le Bourget, the site of the **Paris Air Show.** It also hosts the **Paris Air and Space Museum** created in 1975. Spend a day exploring it before the air show. You'll be glad you did. It presents an outstanding collection of more than 150 aircraft as well as numerous space flight objects. Naturally, examples of the Ariane rocket are shown. You may tour it every day from 10:00 a.m. to 6:00 p.m. For more information visit their Website at http://www.mae.org or e-mail them at musee.air@wanadoo.fr.

The **Paris Air Show** is not open to the public every day, so check well in advance to make certain that you plan to attend on the proper day. Normally the public is invited on the first day and the last two days of the show. The gate opens each day at 9:30 a.m. and closes promptly at 6:00 p.m.

Plan to stay in Paris. It is beautiful, filled with history and great restaurants. Getting around Paris is never easy by car. For Parisians, driving is a blood sport. Always use the train (RER) and the metro. They are generally safe; however, late at night you'll prefer a cab. For your trips to Le Bourget take the metro to Gare du Nord, it is the station for RER B, which gets you close to Le Bourget. You'll have to take a cab from the RER station to the museum and later the air show.

Dates

Odd-numbered years only June 19–20, 2005

Hours

Exhibits	9:30 a.m. to 6:00 p.m.
Flying displays	12:00 p.m. and 5:00 p.m. (narrated in French and English)

Admission Cost

General public	€10 (inc. taxes)
Trade visitors	€35 (inc. taxes)

Contact Information

33 (0) 1 41 69 20 61
marketing@salon-du-bourget.fr
www.paris-air-show.com

Best Place to Stay $$$$$

Hotel de Crillon
10 place de la Concorde
75008 Paris
33 (0) 1 44 71 15 00
crillon@crillon.com
http://www.crillon.com

The Hotel de Crillon is situated in the epicenter of Paris on Place de la Concorde, on the side of the American embassy. It's a block from Rue du Faubourg St-Honoré, and the luxury shops of Hermes, Chanel, and Lacroix. Walk the other way and find yourself beside the Champs-Elysées.

The 115 rooms and 45 suites, six of which have a view overlooking the Place de la Concorde, have all been refurbished in Louis XV style. They all feature soundproofing and air-conditioning, together with many other modern features such as individual mini bar, safety boxes, video and films on request, and access to the Internet via television. All rooms have marble bathrooms.

Health and fitness at the Hotel de Crillon combines the high performance of the most up-to-date equipment with elegant surroundings, offering guests the ultimate in fitness and relaxation. Situated on the sixth floor of the hotel, with exquisite wood fittings, the fitness center is reserved exclusively for the use of guests staying at the Hotel de Crillon.

Best Place to Stay $

Comfort Hotel Gare Du Nord
33, rue de Saint Quentin
Paris, FR, 75 010
(33) 1 4878 0292
(33) 1 4526 8831 (fax)

You'll be able to get a small single room here for well under $100 per night. It is very near the train station which makes getting around Paris a snap. The key to low rates in Paris is to book early and always confirm with a fax and request a confirming fax.

#74

Pima Air and Space Museum

The Arizona Aerospace Foundation has done an excellent job at the **Pima Air and Space Museum.** Few realize it but this is the world's largest privately funded aerospace museum. Since opening in May 1976, its collection of aircraft on display has grown from 75 to over 250. They are scattered across 80 of PASM's 150 acres of southern Arizonian desert.

It will take about a day to see everything that is here. Your best bet is to start with a docent-led tour of the exhibits in hangars 1, 3, and 4. It starts and ends in hangar 1. You'll want to show up a little ahead of time as it is only given twice a day, at 10:15 a.m. and 1:15 p.m. Good news! It's included in the general admission fee.

Another freebie is the **Morphis Movie Ride,** the most unusual simulator you will ever see. It is capable of presenting many different simulations and is changed frequently. I think of it as riding inside a movie.

Be warned, the outside exhibits will drawn you in. You can walk up to each of them for a feel of their skin and an intense study of their markings. There are two problems. First, you are in the middle of the desert. Second, the exhibits sprawl over 80 acres. That is a lot of walking. An hour-long, narrated tram tour departs four times a day, at 10:00 a.m., 11:30 a.m., 1:30 p.m., and 3:00 p.m. It will set you back an extra four bucks per person. *Do it*! You'll become familiar with the entire exhibit field. You can then walk back on your own for an up close look at your favorite ships.

Throughout the year many special events take place at **Pima Air and Space Museum.** Check their Website for the schedule. In April of 2004 they are planning on a visit from the survivors of the Doolittle Raid. A book signing is planned for early in the day with a dinner in the evening.

Dates

Every day of the year except Thanksgiving and Christmas

Hours

9:00 a.m. to 5:00 p.m.

Admission Cost

Adults (ages 13 and up)	$9.75
Seniors, groups of 20 or more, military	$8.75
Children ages 7–12	$6.00
Children 6 and under	Free

Contact Information

Pima Air and Space Museum
6000 East Valencia Road
Tucson, AZ 85706
John Lundquist
(520) 574-0462
(520) 574-9238
http://www.pimaair.org

G/A Airport Serving the Museum

Tucson, AZ (Tucson International—TUS)
Phone: (520) 573-8100
Website: www.tucsonairport.org
Tower: 118.3
ASOS: (520) 889-7236
Runways:

11R–29L	11,000′ × 150′
11L–29R	8,400′ × 75′
3–21	7,000′ × 150′

FBO Name/Phone Number

Tucson Executive Terminal
(520) 573-8128

Ground Transportation

Enterprise Rent A Car
(800) Rent A Car

Best Place to Stay $$$$$

The Westin La Paloma Resort and Spa
3800 East Sunrise Drive
Tucson, AZ 85718
(520) 742-6000
http://www.westinlapalomaresort.com/

It is approximately 20 miles from the museum and well worth the trip. North of Tucson in the lush Sonora Desert, nestled among the foothills of the Santa Catalina Mountains, The Westin La Paloma Resort and Spa offers luxurious guest rooms, a myriad of recreational activities, and the superb service expected of a AAA-Four-Diamond-Award resort. It's easy to see why it was named to the 2002 Condé Nast Traveler Gold List.

Best Place to Stay $

Best Western Inn at the Airport
7060 S. Tucson Boulevard
Tucson, AZ 85706
(520) 746-0271

Conveniently located adjacent to Tucson International Airport is the Best Western Inn at the Airport. This is a great choice for four reasons:

- You can't get any closer to the airport and they run a free shuttle.
- It is only 5 miles to the museum.
- This property has a pool and a putting green.
- The value is exceptional! You should be able to get a room for under $50.

#75

Pioneer Airport

A short tram ride from the Oshkosh EAA AirVenture Museum takes you to **Pioneer Airport,** a unique re-creation of what airports were like during the early years of air travel. You can actually grab a ride on America's first airliner, the Ford Tri-Motor. More than 50 vintage airplanes are displayed in seven period hangars designed to take you back to revive the "Golden Age" of the 1920s and '30s. This is a real grass strip, tube and fabric kinda' place. I love it! So will you.

Here, not only can you see and touch airplanes, you can ride on airplanes, the ones you've seen in countless old photographs and period movies. Do this as soon as you can. These airplanes aren't being manufactured anymore. Even if they were, this is the only authentic airport you can fly one out of. It just doesn't feel the same to fly off of a 1000-foot concrete slab.

Flights take place from **Pioneer Airport's** grass runway every day during the season. You may also see demonstration flights from planes such as EAA's replica of Charles Lindbergh's *Spirit of St. Louis* and the stunning Pitcairn Mailwing biplane once flown by Howard Hughes. All flights are subject to aircraft availability. Call the Pioneer Airport Office for updated information. Buy your flight ticket during your visit to **Pioneer Airport.** Passengers are typically assigned to the next flight. Waiting times are very rarely more than 45 minutes and are often instantly available.

Local flights are available in the following aircraft:

1929 Ford Tri-Motor (My Favorite—Don't miss out!)
The first mass-produced airliner
Seats up to nine passengers
$30 for adults; $25 for children (2003)

1929 Travel Air E-4000 (I'm not a wind in your hair kinda' guy. Are you?)
A classic open-cockpit biplane.
Seats one passenger, $50 (2003)

1927 Spirit of St. Louis Replica
Seats one passenger, $100 (2003)
Pilots can enter this flight in their logbook. That's pretty cool!
This airplane flies under Lindbergh's original N-X-211 registration.

Dates

Open weekdays: Memorial Day to Labor Day
Weekends: First of May through mid-October
Closed New Year's Day, Easter Sunday, Thanksgiving Day, and Christmas Day

Hours

Monday–Saturday	8:30 a.m.–5:00 p.m.
Sunday	10:00 a.m.–5:00 p.m.

Admission Cost

Adults	$8.50
Seniors	$7.50
Students (8–17)	$6.50
Children (7 and under)	Free
Family rate	$21.00

Contact Information

EAA AirVenture Museum
PO Box 3065
Oshkosh, WI
54903-3065
(920) 426-4867
selliott@eaa.org
http://www.airventuremuseum.org/flightops/pioneerairport/

G/A Airports Serving the Event

Oshkosh, WI (Wittman Regional Airport—OSH)

Wittman is a truly wonderful general aviation airport. It offers four runways, the longest of which is 8002 feet. Every IFR approach including an ILS is available.

The museum makes Oshkosh an interesting and convenient fly-in destination, year round. Free shuttle service is available during museum operating hours by parking at Orion Flight Services, Inc.

FBO Name/Phone Number

Orion Flight Services, Inc. (OSH)
(866) 359-6746

Ground Transportation

City	Rental Car Company	Phone
Oshkosh	Hertz Rent-A-Car	(800)654-3131
	Avis Rent-A-Car	(920)730-7575
Appleton	Hertz Rent-A-Car	(800)654-3131
	National Car Rental	(920)739-6421
	Enterprise Rent-A-Car	(800)RENT-A-CAR

Best Place to Stay Oshkosh $$$

Hilton Garden Inn Oshkosh
1355 West 20th Avenue
Oshkosh, WI 54902
(920) 966-1300
(920) 966-1305 (fax)

This premier hotel is right on the Wittman Regional Airport grounds next to the **EAA AirVenture Museum.** Plan on spending 2 days here. You'll need at least that much time!

Best Place to Stay Appleton $$$

Hilton Garden Inn Appleton

720 Eisenhower Drive
Kimberly, WI 54136
(920) 730-1900
(920) 734-7565 (fax)

#76

Piper Aviation Museum

The **Piper Aviation Museum** archives the rich history of William T. Piper, Sr. and his legendary aircraft company. Piper, often called the *Henry Ford of Aviation,* believed that a simple-to-operate low-cost private airplane would flourish, even in the darkest depths of the Great Depression. He was right. By the end of World War II, 80 percent of all U.S. pilots had received their initial flight training in a Piper product, the J-3 Cub. It had become synonymous with general aviation.

In 1931 William T. Piper, an oil man, purchased the Taylor Aircraft Company located in Bradford, Pennsylvania, for $761. The company relocated to Lock Haven, Pennsylvania, purchased a much larger facility and changed the name to Piper Aircraft Company.

Mr. Piper developed an assembly-line production system which enabled the company to become the major small-plane manufacturer with its legendary Piper Cub. Easy to fly and inexpensive, the Cub became known as *the aircraft that taught the world to fly.* The U.S. military was so impressed that it bought thousands that were used for training pilots, artillery spotting, reconnaissance, and many other duties during World War II.

After the war, Mr. Piper, along with his three sons, built more airplanes than anyone else in the world. Piper sold planes in every corner of the globe. At its zenith, Piper Aircraft was operating six manufacturing facilities: three in Pennsylvania, two in Florida, and one in California.

The **Piper Aviation Museum** occupies the former Piper Engineering Center building next to the old Piper factory. Thousands of Pipers were designed and built here from 1937 until the factory closed in 1984. The present museum facility is large enough to accommodate its collections of aircraft, equipment, and various Piper artifacts.

Once a year, Piper aircraft from everywhere return as though Lock Haven were Capistrano and they were swallows. The next "Sentimental Journey" Piper fly-in is scheduled for June 16–19, 2004. The theme of this year's fly-in will be the PA-20, PA-22 Pacer Family. Many pilot attendees will choose to camp out under the wings of their aircraft. There are many good hotels close at hand for those who require more than spartan accommodations.

Interested parties should contact:

Calvin J. Arter, President
Sentimental Journey Inc.
PO Box J-3
Lock Haven, PA 17745-0496

(570) 893-4200 or (570) 893-4207
(570) 893-4218 (fax)
j3cub@kcnet.org
www.sentimentaljourneyfly-in.com/

Dates

The museum is open every day of the year. It is closed on Sundays during winter until May 4.

Hours

Monday–Friday	9 a.m. to 4 p.m.
Saturday	10 a.m. to 4 p.m.
Sunday	12 noon to 4 p.m.

Admission Cost

Adults	$5
Seniors	$4
Children 6–12	$1
Under 6	Free

Website

www.pipermuseum.com

Contact Information

Piper Aviation Museum
One Piper Way
Lock Haven, PA 17745
(570) 748-8283
piper@cub.kcnet.org

G/A Airport Serving the Event

Lock Haven, PA (W.T. Piper Memorial Airport—LHV)
Phone: (570) 748-5123
CTAF-Unicom: 122.8
Runway: 9–27 3824′ × 100′

FBO Name/Phone Number

City of Lock Haven
(570) 748-5123
They get a lot of Cubs so they actually offer 80 octane fuel!

Ground Transportation

Enterprise Rent-A-Car
(800) Rent A Car

Best Place to Stay $

Best Western Lock Haven
101 East Walnut Street
Lock Haven, PA 17745
(570) 748-3297

The Best Western Lock Haven is only 1 mile from the airport and offers many simple pleasures for the value-minded travelers of Pennsylvania. This 67-room property strives to provide first-class service with clean and comfortable accommodations to all guests. Deluxe complimentary continental breakfast, laundry facilities, and a fitness center are included.

Best Aviation Attraction

#77

Planes of Fame

The **Planes of Fame** museum collection spans the history of manned flight from the Chanute Hang Glider of 1896 to the space age. Of particular pride to the museum is its collection of Japanese aircraft, which is currently the largest of its type in the world. This collection includes the world's only totally authentic flying Japanese Mitsubishi "Zero" fighter, which is complete with its original engine.

Mr. Edward Maloney, the founder, recognized the importance of preserving World War II aircraft at a time when most of these planes were being cut up into scrap metal. The United States produced over 300,000 military aircraft during World War II, but most of these were destroyed at the end of the war and many types disappeared entirely. Even fewer of Germany's and Japan's aircraft were saved. In his attempts to save endangered aircraft types from the scrap heap, Mr. Maloney pleaded, bartered, and even purchased the discards by the pound.

A number of the aircraft on display at the museum are *sole surviving examples* of their type and still exist only because of Mr. Maloney's personal determination to save at least one example of as many different aircraft as possible.

Parts for the aircraft have been collected from all over the world, with reclamation efforts ongoing. The museum compound even houses a full-time warbird restoration facility called *Fighter Rebuilders.* This facility allows the museum to have several restoration projects going on simultaneously. Of the museum's approximate 150 aircraft, 30 are flyable. On a typical Saturday, you may see two P51 Mustangs fly by escorting a B-25 Mitchell bomber or a Grumman Hellcat with a Chance-Vought Corsair making a formation overhead approach to the airport.

The **Planes of Fame** was the first permanent air museum west of the Rocky Mountains. It officially opened its doors to the public in January 1957, with an initial collection of six aircraft and a great deal of hope for the future. The museum's original location was in Claremont, California. As the collection began to outgrow this first makeshift facility, the aircraft were moved to the Ontario Airport in California. In 1973, the museum finally took up its present residence at the Chino Airport, California. Coincidentally, this location was originally the home of the Cal-Aero Flight Academy during World War II, where thousands of Army Air Corps cadets learned to fly the planes the museum now preserves.

This is a "living history" museum, where the aircraft are not only preserved but kept flying. To share their collection with the public, the warbirds are flown regularly, participating in air shows and military base open houses. They are often used in the making of television programs and motion pictures.

The airplanes may have been built in a time gone by, but the unforgettable sound and smell of the aircraft, as well as the enthusiastic spirit of the pilots who flew them to world records, trophies, and combat victories, are still alive and well at **Planes of Fame.**

Orientation flights are available for those individuals who are members or become members and wish to make a special donation to enable the **Planes of Fame Air Museum** to fly a specific aircraft. The member may then choose to watch from the ground or go along on the flight!

The following aircraft are available for orientation flights: P-51 Mustang, P-40 Warhawk, SBD Dauntless, B-25 Mitchell Bomber, and T-6 Texan.

Dates

Open every day of the year except Thanksgiving and Christmas

Hours

9:00 a.m. to 5:00 p.m. daily

Admission Cost

Adult admission	$8.95
Children 5–12	$1.95
Children under 5	Free

Website

http://www.planesoffame.org/

Telephone Number

(909) 597-3722

G/A Airport Serving the Event

Chino, CA (Chino Airport—CNO)

FBO Name/Phone Number

Jet Executive Transportation Technologies
(800) 720-5388
Chino Air Center
(909) 597-2990
Chino Fuel Service, LLC
(909) 393-0782

Ground Transportation

Hertz
(909) 482-1835

Best Place to Stay $$$

Sheraton Ontario Airport Hotel
429 N. Vineyard Avenue
Ontario, CA

Best Place to Stay $

Econo Lodge Airport South
2301 South Euclid Avenue
Ontario, CA 91762

#78
Prairie Aviation Museum

Prairie Aviation Museum is located in Bloomington, Illinois, at the old historic terminal of the Central Illinois Regional Airport. It features permanent and rotating exhibits, a minitheater, and static outdoor displays of historical aircraft and vehicles. At 2400 square feet to say the museum is small is to be generous. The reason to come here is in a nearby hangar.

It is a 60-year-old fully operational airliner, N763A, Douglas DC-3 Ser. 4894 to be exact. Today she sports the colorful green and white paint job of Ozark Airlines, where she once flew! On a crisp October afternoon in 1998, N763A became the first DC-3 to be listed on the National Register of Historic Places. At the same time she became the only flying aircraft to be listed.

A visit here will be your best chance to grab a ride on the most remembered airliner of commercial aviation history, the DC-3! A DC-3 **Discovery Flight** can be yours for as little as $80. That, my friend, is a great bargain; however, my advice is to spend just a little more and buy a ride in the cockpit jump seat for $105. You'll get a bird's-eye view of what the crew sees and does to get this bird in the air and bring it back to earth after your trip is completed.

Each flight operates on a first-come, first-served basis, and the nine first-class seats fill up quickly. The jump seat even faster! Your **Discovery Flight** will last about a half hour from engine start to shutdown; of that time 20 minutes will be in the air. Passengers are treated to a ground tour before and after each flight.

If you're a pilot, you'll be delighted to know that flight training is offered in this aircraft. You can pick up an IFR or VFR rating, co-pilot checkout, or a FAR 61.58 recurrency check. Yes, it is expensive. Flight training in this ship goes for $1200 per flight hour. A well-trained pilot can pass the DC-3 IFR checkride in about 6 hours. The VFR course is normally completed in 3 hours.

Dates

Open daily
Closed Thanksgiving, Christmas, and New Year's Day

Hours

Tuesday–Saturday	11:00 a.m. to 4:00 p.m.
Monday	Closed
Sunday	12:00 to 4:00 p.m.

Admission Cost

Adults	$2
Children (6–11)	$1
Children (5 and under)	Free
Members	Free

Contact Information

Prairie Aviation Museum
2929 East Empire Street
Bloomington, IL 61701
(309) 663-7632
(309) 663-8411
www.prairieaviationmuseum.org
info@prairieaviationmuseum.org

G/A Airport Serving the Event

Bloomington, IL (Central IL Regional Airport—BMI)
Phone: (309) 663-7383
Website: www.cira.com
Tower: 124.6
Runways:

11–29	6500′ × 150′
2–20	7000′ × 100′

FBO Name/Phone Number

Image Air
(800) 232-4360
www.imageair.com

Ground Transportation

Rental car

Best Place to Stay $$$

Radisson Hotel and Conference Center
10 Brickyard Drive
Bloomington, IL 61701
(309) 664-6446

If you decide to fly in commercial or private, this is a good hotel at the airport with shuttle service back and forth.

#79

St. Louis County Fair & Air Show

St. Louis County Fair & Air Show is near the top of my mid-America air show list. These folks have added a whole new twist to the flight adventure weekend. They've combined a slick air show with a county fair. Well not a county fair really but a fair just the same. Included are world-class carnival rides, a first-rate air show, the area's largest static aircraft display, Anheuser Busch's world famous Clydesdales, Faust Park Historical Village, Bud World, Kids Town, Purina Farm's petting zoo, Purina's Incredible Dog Team, and the Show-Me Missouri Fish Mobile Aquarium. Want more? Saturday and Sunday nights get really cranked up when a fireworks display follows the concerts on the main stage. One more thing: the lead air show act is normally the Blue Angels! This is a first-class entertainment weekend from start to finish. Do not miss a minute of it.

So here's what they've got: air show, fair, concert, and fireworks. If you live or travel in the heartland of America you need to go to this one. It is a terrific weekend.

Dates

September 4–6

Hours

Friday	5:00 p.m.–10:00 p.m.
Saturday	10:00 a.m.–10:00 p.m.
Sunday	10:00 a.m.–10:00 p.m.
Monday	10:00 a.m.–6:00 p.m.

Admission Cost

Adult	$7
Seniors	$3
Child (6–12)	$3
Child (5 and under)	Free

Contact Information

St. Louis County Fair & Air Show
18612 Olive Street Road
Chesterfield, MO 63005

(636) 530-9386
http://www.stlcofair.org/

G/A Airport Serving the Event

Chesterfield, MO (Spirit of St. Louis Airport—SUS)
Phone: (636) 532-2222
Tower: 124.75
ASOS: (636) 536-0159
Runways:

8L–26R	5000′ × 75′
8R–26L	7004′ × 150′

FBO Name/Phone Number

Million Air St. Louis
(636) 532-0404

Ground Transportation

Rental car

Best Place to Stay $$$

Doubletree Hotel & Conference Center St. Louis
16625 Swingley Ridge Road
Chesterfield, MO 63017-1798
(636) 532-5000

Best Place to Stay $$$

Marriott—St. Louis West
660 Maryville Centre Drive
St. Louis, MO 63141
(314) 878-2747

Best Aviation Attraction

#80

San Diego Aerospace Museum

The **San Diego Aerospace Museum** was the first aviation museum to be accredited by the American Association of Museums, and it continues to be one of the top five aviation and space museums in the nation. The San Diego community has long been at the forefront of advances in the world of aviation. Lindbergh came to San Diego to have the Ryan Company build the *Spirit of St. Louis.* It was from San Diego that his flight into history began.

There is much that can be seen here that can be seen nowhere else. In the rotunda a replica of the *Spirit of St. Louis* is displayed before you on the floor so you may get a good glimpse of it. Travel to the Smithsonian and your gaze will be interrupted by the distance between you and the plane suspended from the ceiling. What is special about this one replica when there are so many? This one was built by the same team of Ryan employees who built the original. The **International Aerospace Hall of Fame** is housed here. It is the world's only attempt to honor *all* of the men and women who made substantial contributions to the advancement of the aerospace sciences. Lindbergh, Montgolfier, Earhart, Gagarin, and Armstrong and many others are remembered; all are honored.

The secret to enjoying and learning from this museum is found in the city that hosts it. San Diego is an aviation center. Primarily it is a naval aviation center. Look across the bay toward Coronado Island. It is hard to not notice the naval aviation facility on the northern end. North Island, as it is called, is one of the most impressive naval aviation facilities in the word. Often three aircraft carriers will be in port here at the same time, and there are times that you can tour them. These opportunities change often and are very dependent on the particular ship that is in port and our nation's security posture.

Dates

Open daily

Closed Thanksgiving Day, Christmas Day, and New Year's Day

Hours

Daily 10:00 a.m.–4:30 p.m.

Admission Cost

Adults	$8
Juniors (6–17 years)	$3

Seniors (65+ years)	$6
Students (with valid student ID)	$6
Children (under 6)	Free

Contact Information

San Diego Aerospace Museum
2001 Pan American Plaza
Balboa Park
San Diego, CA 92101
(619) 234-8291
information@sdasm.org
http://www.aerospacemuseum.org/

G/A Airport Serving the Event

San Diego, CA (Gillespie Field—SEE)
Phone: (619) 596-3900
CTAF-Unicom: 120.7
AWOS: (619) 448-1641
Runways:

17–35	4147′ × 100′
9R–27L	2737′ × 60′
9L–27R	5341′ × 100′

Gillespie Field is the home of the **Museum's Annex.** Here aircraft are not only restored to new condition but often built from scratch. The museum's education wing has an impressive learning center here which is used to educate local school children on all aspects of aviation.

FBO Name/Phone Number

Royal Jet
(619) 448-4200

Ground Transportation

Rental car

Best Place to Stay $$$$$

Hotel Del Coronado
1500 Orange Avenue
Coronado, CA 92118
(800) 582-2595
http://www.hoteldel.com/

For me there is only *one* hotel to stay at when I'm in San Diego. It is the Del Coronado. Many, many movies have been filmed here. It is a unique hotel to be sure. Its main attraction though is its location. The Del is located on what many consider to be one of the 10 best beaches in America. What makes the beach even more interesting are the other beach goers that you'll see here and nowhere else. The Navy trains its SEALS at its Coronado Base, which is adjacent to the hotel's property. It is not unusual at all to see them training on the same beach where you'll be enjoying a hotel-provided hamburger!

Life's contrasts are amazing, aren't they?

#81
Seafair

Seafair is Seattle's traditional month-long summer festival that brings the entire region together. This summer will be time number 55! The evergreen city will be sprinkled with community events, parades, amateur athletics, boat racing, and of course an air show!

In 2003 more than 250,000 spectators lined the beaches of Lake Washington watching spectacular shows on the water and in the air: the 2003 GM Cup (a hydroplane race) and the KeyBank Air Show. In 2004 the boat race will have a name change and become the Chevrolet Cup. Think about this scene for a moment, hydroplane racing and the Blue Angels at the same event!

The air show is best seen from a boat anyplace you can anchor on Lake Washington. If you can't work that out, find a good picnic spot at any of the shoreline parks. When No. 6 makes its low pass over the lake the water literally separates in its turbulence. It is worth seeing. As always, my advice is to get set up for practice day. You'll be able to get some great photos without the pressure of the crowd.

Seafair is amazing. I suppose it is a part of the reason that Northwesterners love Puget Sound.

Dates

August 6–8, 2004

Hours

Air show starts at 10:00 a.m. with the Blues performing at 3:00 p.m.

Admission Cost

Free

Contact Information

Seafair
2200 Sixth Avenue, Suite 400
Seattle, WA 98121
(206) 728-0123
info@SEAFAIR.com
http://www.seafair.com/

G/A Airport Serving the Event

Seattle, WA (Boeing Field/King County International—BFI)

Phone: (206) 296-7380

Website: www.metrokc.gov/airport/

Tower: 120.6

ASOS: (206) 763-6904

Runways:

13L–31R	3710′ × 100′
13R–31L	10,000′ × 200′

FBO Name/Phone Number

Galvin Flying Service, Inc.

(206) 763-0350

(206) 767-9333 (fax)

Ground Transportation

Rental car

Best Place to Stay $$$

Red-Lion Seattle-South

11244 Pacific Highway South Seattle

Seattle, WA 98168

Phone: (206) 762-0300

Fax: (206) 762-8306

Res: (800) RED-LION

sales@redlionseattle.com

This 3-diamond hotel has all the amenities of a true business-class hotel. It provides complimentary airport shuttle service to Boeing Field, SeaTac Airport, the Museum of Flight, and limited service to Southcenter Mall and downtown Seattle, including the Amtrak train station.

#82

Smithsonian National Air and Space Museum

The Smithsonian Institution's **National Air and Space Museum** is thought to maintain the largest collection of historic airplanes and spacecraft in the world. It is currently divided into two display facilities. Both are in our nation's capital. The original building is located on the National Mall near the more prominent Smithsonian facility. Here you will find hundreds of artifacts including a Wright Flyer attributed to1903. As we all know, the "original" was severely damaged following its fourth flight on December 17, 1903. It was never repaired and I am almost certain that it was the only one the Wrights had at that point. They left Kitty Hawk shortly after December 17, 1903, so they could be home in Ohio for Christmas. I don't think they had enough time following their first flight of December 17 to build another one in 1903.

Lindbergh's *Spirit of St. Louis* hangs from the ceiling here. It is good to be in the same room with it but sad that it cannot be approached. The Apollo 11 command module is displayed as is a hunk of Lunar rock.

Way across town, or should I say out of town, is the new **Steven F. Udvar-Hazy Center.** It is actually not in Washington at all but in nearby Chantilly, Virginia, near the Dulles International Airport. One of several copies of the Lockheed SR-71 Blackbird is housed here as is the Space Shuttle Enterprise. Finally, the Boeing B-29 Superfortress Enola Gay is here also. There was much controversy spanning many years as to just how it would be displayed and what words would be printed on the exhibitory. The politically correct crowd really came out of the woodwork for this one. I haven't seen it yet so I can't comment on how it turned out, but I am hopeful.

Bob Hoover's air show-famous twin-engine Commander is also here and should be. His air show performances in that plane were the best I ever saw!

There is much to see, do, read, and study at either of these facilities. How much time should you allocate for your visit? How much time do you have is a better question for there is truly that much here.

If you can go to only one museum this year I would probably not pick this one, but if you could go to two this is a must. Certainly, if you are curious about airplanes and the people who design, build, and fly them you must come here at least once in your life. There is no time like the present.

The Washington Mall is one of the easiest places to get to in the country. If you're East Coast–based take the Amtrak. It will let you off at Union Station which is a very short walk from the Smithsonian's mall-located complex.

If you are flying commercial, opt for Washington's convenient Ronald Reagan Washington National Airport. Savvy travelers know to take the metro from here directly to the mall. Cabs are available as are rental cars. Travel light and take the metro.

The problem comes if you elect to fly yourself. Our capital is today covered with TFRs and the three most convenient general aviation airports are for all practical purposes just not available.

Dates

Open daily
Closed Christmas Day

Hours

10:00 a.m.–5:30 p.m.

Admission Cost

Free

Contact Information

Smithsonian National Air and Space Museum
4th and Independence Avenue, SW
Washington, DC 20560
(202) 357-2700
info@si.edu

G/A Airport Serving the Event

Due to the nature of the TFRs around Washington, transient private aircraft are not permitted at any convenient airport.

Best Place to Stay $$$$$

Willard
1401 Pennsylvania Avenue NW
Washington, DC 20004
(202) 628-9100
washington@interconti.com

I kid you not, this is the place to stay. It is really convenient and they have a terrific staff. You will want for nothing, just ask. They are *very* wired and can get you tickets to anything.

#83

Southern Wisconsin AirFEST

The **Southern Wisconsin AirFEST** has the most spectacular twilight air show I have seen. It starts at 4:30 p.m. when seven performers take to the twilight. Then as darkness falls the Aeroshell Team provides a night performance. The thing you're here to see is Shockwave, the jet-powered super truck. On a dark night this ground shaker will rock your socks off as it zooms down the runway trailed by the bright fire of its jet plume. Show management promises that the Friday night event will bring "ooh's, ahh's and even tears" as it concludes with AirBooms, the most spectacular and moving fireworks display you've ever seen. So they say. We'll see!

Friday night performers will be Masters of Disaster, Super Shockwave Jet Truck, Aeroshell Team, Jet Powered Biplane, Flight for Diabetes, and "Big Red" Stearman.

This is not just a one-night show nor is it a one-trick pony. Saturday and Sunday are power-packed, especially this year with the return of the Thunderbirds. Remember just a couple of things to make the show perfect for you. It is going to be hot and the sun is going to be bright. Bring sunscreen and sunglasses. The best sunglasses have polarized lens, spend a few dollars more and make sure yours do. Don't mess around with your eyes; just one pair to a customer!

Security is tight at all crowd-pulling events these days, especially air shows. Arrive early and count on being searched. Make it easy on yourself and everyone else by bringing in as little as possible. We all know that it is no longer possible to enter with a cooler or chairs in most cases.

There will be plenty of military gear flying at this show. I wouldn't be surprised if a B-2 did a flyby. I have seen them often in this part of the country.

Dates

June 25–27, 2004

Hours

Friday

Gates open	3:30 p.m.
Show begins	4:30 p.m.
Fireworks	7:30 p.m.

Saturday and Sunday

Gates open	9:00 a.m.
Show begins	12:30 p.m.
Thunderbirds	3:30 p.m.
Gates close	5:30 p.m.

Admission Cost

Adult	$15
Youth (6–12)	$10
Child	Free

Contact Information

WAA—SW AirFEST Office
4606 South Atlantis Drive
Janesville, WI 53546
(608) 373-0904
(608) 373-0906 (fax)
info@SWAirFEST.org

G/A Airport Serving the Event

Janesville, WI (Rock County—JVL)
Phone: (608) 757-5768
Tower: 118.3
ASOS: (608) 758-1723
Runways:

14–32	7300′ × 150′
4–22	6701′ × 150′
18–36	5000′ × 75′

Incoming aircraft should arrive before 1:30 p.m. on Friday, and before 12:00 noon on Saturday and Sunday.

FBO Name/Phone Number

Janesville Jet Center
(608) 758-1037

Ground Transportation

Rental car

Best Place to Stay $$

Ramada Inn
3431 Milton Avenue
Janesville, WI 53546
(608) 756-2341

Best Place to Stay $$

Baymont Inn and Suites
616 Midland Road
Janesville, WI 53546
(608) 758-4545

#84

Space Camp

Space Camp, the nation's leading science-based camp program for children, has been operating since 1982. It is also the largest camp operation in the United States, having graduated almost 400,000 campers. **Space Camp** programs in Alabama are operated by the U.S. Space & Rocket Center and the Alabama Space Science Exhibit Commission.

The safety of children is a top priority. **Space Camp** is located within a completely enclosed facility. Security guards, on duty around the clock, routinely check all packages and parcels entering the facility, even those of the employees. More than 50 video cameras constantly monitor the facility.

There are many camping options to choose from even though the original **Space Camp** package is still the most popular. It is a 4-day program packed with astronaut training for young people. Activities include simulated space shuttle missions, IMAX movies, training simulators (like the 1/6th Gravity Chair), rocket building and launches, scientific experiments, and lectures on the past, present, and future of space exploration.

To qualify, children must already be attending classes in the fourth grade and be at least 9 years old. Trainees are housed in the Space Habitat, a futuristic space station mock-up, featuring individual rooms of up to seven persons, or bays of 20–40 beds. The Habitat complex accommodates 800 people at full capacity. Bed sheets, blanket, pillow, and pillowcase are provided. Towels and washcloths are not provided. Bring a beach towel for water activities. Each trainee will have an assigned locker. The $499 tuition includes all meals, lodging, and program materials. It does not cover transportation.

When my daughter is old enough, children must be 7–12, we're going to do **Aviation Challenge.** It is a parent–child program scheduled over a 3-day weekend. The program combines simulated jet fighter pilot training, with guided tours of some of the finest flying machines ever built. We'll also do missions together, side by side, as teammates.

Together, we'll practice land and water survival techniques. Where's the escape route from the Helo Dunker? This builds confidence by encouraging trainees to try new things, meet new people, and overcome challenges in a safe and positive environment.

The $698/pair tuition includes everything but transportation. Naturally, we'll fly our Cherokee up from Orlando. For $898 we can select the Parent/Child Plus 1 program and bring mom along.

Dates

Open weekly during the summer

Hours

Daily, 24 hours a day

Admission Cost

There are many programs available. They are priced separately.

Contact Information

Space Camp
Huntsville, AL
(800) 63 SPACE
http://www.spacecamp.com
guestservices@spacecamp.com

G/A Airport Serving the Event

Huntsville, AL (Huntsville International—HSV)
Phone: (256) 772-9395
Tower: 127.6
ASOS: (256) 772-8074
Runways:

18R–36L	8000′ × 150′
18L–36R	10,000′ × 150′

FBO Name/Phone Number

Signature Flight Support
(256) 772-9341

Ground Transportation

Rental car

Best Place to Stay $$$

Radisson Suite Hotel Huntsville
6000 Memorial Parkway South
Huntsville, AL 35802
(256) 882-9400

#85

Space Center Houston

Space Center Houston is the official visitors center of **NASA's Johnson Space Center,** which is the home of astronaut training and mission control. JSC and NASA have a tremendous story to tell. This is not just a space flight center but a manned space flight center. All of the history of the United States efforts to put man into space happened here. The original seven Mercury astronauts worked here and lived in the neighborhoods just outside the complex. John Glenn's historic three-orbit flight was controlled from Building 30 on this site as was Alan Sheppard's 15-minute ride which put America into the business of flying men into space. Later Neil Armstrong's landing at Tranquility Base would be shown to the world from here. "Houston we've got a problem" and "failure is not an option" are but a few of the memorable words which punctuated the historic moments that played out behind these walls. In a very real way JSC is the Kitty Hawk of the Space Age. This is not a place of history as much as it is the launch pad of tomorrow.

Space Center Houston was brought to life to tell the story of man's greatest adventure, our travel to the stars. The experts from Walt Disney Imagineering, the design and master planning arm of the Walt Disney Co., generated the concepts that would become **Space Center Houston.** BRC Imagination Arts produced the shows and displays the Disney team envisioned. The center entertains and excites, while telling the true story of space flight.

Your visit here should begin at the Starship Gallery; the film *On Human Destiny* at the Destiny Theater is a good first stop. This incredible artifact collection includes an original model of the Goddard Rocket, the actual Mercury Atlas 9 "Faith 7" capsule flown by Gordon Cooper, the Gemini V Spacecraft piloted by Pete Conrad and Gordon Cooper, a Lunar Roving Vehicle Trainer, the Apollo 17 Command Module, the giant Skylab Trainer, and the Apollo-Soyuz Trainer. It is a lot to see, but it is only a small part of what is to be seen.

The brand-new Blastoff Theater visit helps you to personally experience the thrill of launching into space. It's more than a movie. It's like IMAX in 10D! There's nothing like it in the whole world!

Do not forgo the Level Nine Tour. It takes you into the work-a-day world of NASA. For 4 hours you go where only the astronauts go, see what only they see, and eat where they eat. A knowledgeable tour guide will aid your discovery of the secrets that have been kept behind these closed doors. Level Nine Tour is not offered on week-ends and you must be 16 years of age or older. Only 12 tickets are available each day. Advanced reservations must be made at least a day prior to visit. You'll vist the following NASA buildings:

Building 3—Astronaut cafeteria (Enjoy a free lunch where the astronauts eat everyday.)

Building 2—Teague auditorium (artifacts)

Building 7—Space suit displays
Building 9NE—Space vehicle mock-up facility (observation catwalk)
Building 30S—New mission control center (viewing areas)
Building 30N—Historic mission control center (I once worked here!)
Building 32—Space environment simulation lab (vacuum chamber)
Sonny Carter Training Facility—Neutral buoyancy lab (observation catwalk)

Enjoy your visit. NASA–JSC is the place where man's grasp is aligned with his reach. It is the place where dreams come true. We've been to the moon. It's time to go back and beyond.

Dates

Open daily

Hours

Summer:

June	10:00 a.m.–7:00 p.m.
July	9:00 a.m.–7:00 p.m.
August	10:00 a.m.–5:00 p.m.
Weekends	10:00 a.m.–7:00 p.m.

Winter:

Monday–Friday	10:00 a.m.–5:00 p.m.
Weekends	10:00 a.m.–7:00 p.m.

Admission Cost

Adults	$17.95
Children (4–11)	$13.95
Seniors	$16.95
Parking	$4.00

Contact Information

Space Center Houston
1601 NASA Road 1
Houston, TX 77058
(281) 244-2100
http://www.spacecenter.org

G/A Airport Serving the Event

Houston, TX (Ellington Field—EFD)
Phone: (713) 847-4200
Tower: 126.05
Runways:

17L–35R	4609′ × 75′
17R–35L	9001′ × 150′
4–22	8001′ × 150′

FBO Name/Phone Number

Southwest Airport Services, Inc.
(281) 484-6551

Ground Transportation

Rental car

Best Place to Stay $$$

Residence Inn—Clear Lake
525 Bay Area Boulevard
Houston, TX 77058
(281) 486-2424

Best Place to Stay $$$

Hilton—NASA
3000 NASA Road One
Houston, TX 77058-4322
(281) 333-9300

Best Aviation Attraction

#86

Sun 'n Fun

(Courtesy EAA.)

When I think about fly-ins and fly-in conventions my first thought is of Oshkosh and The Big Show that the EAA puts on every year at its Oshkosh, Wisconsin, headquarters. Ironically the first major fly-in of the season is also an EAA event. It takes place 1200 miles to the south of Oshkosh in beautiful Lakeland, Florida. In 2004 it is correctly named *The 30th Anniversary of the Spring Celebration of Flight.* Major milestones always herald stupendous events. If you have never been to **Sun 'n Fun** before, this is the year to go. It is so much more than an air show. It is workshops, seminars, speakers, and the best vendors in the aviation business eager to answer your every question about their latest technology.

Sun 'n Fun typically takes place in early to mid-April. It is now on a Tuesday to Monday schedule; in the past it went from Wednesday to Tuesday. Certainly **Sun 'n Fun** is the second largest fly-in of each year. Like its big northern brother it is somewhat like Woodstock. It is a gathering of the faithful. Pilots from around the world show up early and stay late. Many camp out on the grounds. The EAA has a permanent facility on the grounds of Lakeland's Linder Field (LAL).

Sun 'n Fun has its own Website, which is filled with all the details that you'll need to plan your visit. Make sure your itinerary has Saturday night open. You will want to see the nighttime air show. When the sun goes down, an aerial performance takes on a special luster.

Dates

April 13–19, 2004

Hours

Daily sunup to sundown
Saturday nighttime air show

Admission Cost

EAA members	$25.00
Nonmembers	$30.00
Youth (13–17)	$17.00
Children (12 and under)	Free

Contact Name

Sun 'n Fun
PO Box 7670
Lakeland, FL 33807
(863) 644-2431
http://www.sun-n-fun.org

G/A Airport Serving the Event

Lakeland, FL (Lakeland Linder Regional—LAL)
(863) 648-3299
http://www.lakelandgov.net/airport/home.html
Check NOTAMS. During **Sun 'n Fun** everything concerning operating from this airport is special!

FBO Name/Phone Number

During the show the EAA is the FBO!

Ground Transportation

Rental car

Best Place to Stay $

Sun 'n Fun campgrounds
Lakeland-Linder Field (LAL)
http://www.sun-n-fun.org/

This is the best place to stay during the air show. You won't miss a thing but the traffic and the parking lot madness! The facilities are really very nice. There are showers, toilets, a general store, and numerous eating venues. If you fly in, there are five special areas set aside for you: general aircraft camping, homebuilt aircraft camping, handicapped aircraft camping, overnight aircraft camping, and vintage aircraft camping.

Throughout the year these camping areas are available for $20 per night. During the convention they are $80 for the week! The better deal is to buy a package which includes two show admissions and campground fees for only $215. That saves you about $65!

Best Place to Stay $$$$

The Ritz-Carlton Orlando, Grande Lakes
4012 Central Florida Parkway
Orlando, FL 32837
(407) 206-2400
(407) 206-2401 (fax)
http://www.ritzcarlton.com

Here's the best news about Lakeland, Florida. It is only 40 miles from Orlando with its world-class resorts and unequaled theme park. This hotel and spa opened in 2003. From the moment you walk in you know you're in a Ritz. Reward your spouse with a spa package or a day on the golf course. It is the best way to say thanks for hanging out at an air show with me! For a few dollars less you can stay in the attached Marriott, which is also brand new! Either way you're in for a treat.

#87

TICO Warbird Airshow

There are two very good reasons to come to this event, maybe three. The first is that it is located right next door to the Kennedy Space Center. You could not possibly plan a better weekend than one which includes a trip to an air show featuring propeller-driven warbirds from World War I and World War II. Then move next door and see the hardware that lifts man into space on a regular basis. It is your good fortune that these two venues, one which focuses on a history of 60 and more years ago and one which looks to the future, could be yours in one weekend.

At the **TICO Warbird Airshow** you'll not only see warbirds fly but also dogfights and bomb-and-strafe the field. This is a very cool show. Most of the planes will be vintage World War II. Your best bet is to do NASA KSC on Saturday and the air show on Sunday. Most of the crowd will do just the reverse. Outflank 'em!

On Sunday, your air show day, arrive early, remember you cannot get in with a cooler or a dog. Buy the flight line pass, a must. Bring your camera, the digital one, and take a ton of pictures of the aircraft on the flight line that will later perform in the show. This is a smaller more casual venue than you may be used to, so take advantage of it. Many of the pilots will start getting to their planes well before the 1:00 p.m. air show start time. The flight line doesn't close until noon. If you play a heads-up game, you can snag a pilot or two and ask all of those "what does it feel like to…" questions. They'll be glad to fill you in and to pose for a photo with your children in front of their plane. Flatter them and ask for an autograph, and be sure and have them sign next to their call out in the show program.

Be warned! The road between Orlando and Titusville is narrow and crowded. Decide to either leave early or stay late. Your best bet is to book a beachfront motel or condo at nearby Cocoa Beach and spend an extra day.

Dates

Annually mid-March (Saturday and Sunday)

Hours

Gates open 8:30 a.m.

Air show starts 1:00 p.m.

Admission Cost

Adults	$12.00 gate (advance $10.00)
Children (4–12)	$ 7.00 gate (advance $5.00)
Flight line pit pass	$ 3.00 (9:00 a.m.–12 noon)
2-day pass (includes flight line)	$20.00

Contact Information

TICO Warbird Airshow
US 1 Route 405
Titusville, FL
(321) 268-1941

G/A Airport Serving the Event

Titusville, FL (Space Coast Regional—TIX)
Phone: (321) 267-8780
Tower: 118.9
Runways:

18–36	7320′ × 150′
9–27	5000′ × 100′

The view when you're landing and taking off is amazing. Typically you'll take off on runway 9 which will point you straight toward NASA's Saturn launch gantry, which you will see very clearly just on the other side of the Banana River. You will not reach 500 feet to make your turnout until you are halfway across the river. The phrase *getting there is half the fun* really applies to this trip!

FBO Name/Phone Number

TICO Executive Aviation
(321) 267-8355
www.ticoaviation.com

Ground Transportation

Rental car

Best Place to Stay $$$$$

Holiday Inn—Kennedy Space Center
4951 S. Washington Avenue (US 1)
Titusville, FL 32780
(321) 269-2121
(321) 267-4739 (fax)
HIKSC@bellsouth.net

Best Place to Stay $$

Days Inn
3755 Cheney Highway (Highway 50)
Titusville, FL 32780
(321) 269-4480

Best Aviation Attraction

#88

U.S. Air Force Museum

USAF Museum. (© *USAF Museum.*)

The **U.S. Air Force Museum** near Dayton, Ohio, is the oldest and largest military aviation museum in the world. The museum uses both chronological and subjective layouts to tell the exciting story of aviation's development from the days of the Wright brothers at Kitty Hawk to the Space Age. Exhibits include over 300 aircraft and missiles, plus family-oriented and historically interesting aeronautical displays. Over a million and a half visitors from around the world come to Ohio each year to tour this unique free attraction.

The presidential and research and development flight test hangars, housing approximately 50 aircraft including the museum's collection of presidential aircraft, are located on the active part of Wright-Patterson Air Force Base (WPAFB). They are accessible using the shuttle bus service from the main museum complex. It is free and runs continuously from 10 a.m. to 4 p.m. This service is offered on a first-come, first-served basis. Shuttle buses are not handicapped accessible. Individuals requiring special assistance should contact the Museum Operations Division in advance to arrange transportation at (937) 255-3286. Military and DoD civilians with government ID may use their private vehicles to visit the presidential and R&D hangars.

Audiences can travel on an incredible cinematic journey of discovery to the international space station at the U.S. Air Force Museum's IMAX Theatre. *Space Station* stars astronauts and cosmonauts, from the United States, Canada, Japan, Russia and Europe, who collectively have spent thousands of hours in space, as they construct the first human outpost. The IMAX cameras captured seven space shuttle crews and two resident station crews, as they transformed the international space station into a permanently inhabited scientific research station. The men and women of the crew share the triumphs and tensions of their greatest challenge—hours of painstaking and dangerous teamwork in the deadly vacuum of space—to complete the greatest engineering feat since landing a man on the moon. You can take this cinematic journey daily at 11 a.m. and 3 p.m. For updated information on *Space Station* or other IMAX presentations, please call the USAF Museum's IMAX theater at (937) 253-IMAX or visit their Website at http://afmuseumimax.com/imax/imax.html.

Dates

Open every day except Thanksgiving, Christmas, and New Year's Day

Hours

9:00 a.m. to 5:00 p.m. daily

Admission Cost

Free

Website

http://www.wpafb.af.mil/museum/

Telephone Number

(937) 255-3284

G/A Airports Serving the Event

Dayton, OH (James M. Cox Dayton International Airport—DAY)
Dayton, OH (Greene County-Lewis A. Jackson Regional—I19)

FBO Name/Phone Number

Stevens Aviation (DAY)
(937) 954-3400
Wright Brothers Aero (DAY)
(937) 890-8900
Greene County Airport Authority (I19)
(937) 376-8107

Ground Transportation

Enterprise Rental Cars

Best Place to Stay $$

Holiday Inn
300 Xenia Towne Square
Xenia, OH 45385

(800) 465-4329
(937) 372-9921

The Xenia Holiday is only 2.5 miles from the Greene County Regional Airport. If you're flying in and want a close motel this is a good pick. However, you'll be in for an 11-mile cab ride to the museum.

Best Place to Stay $$

Holiday Inn—Fairborn
2800 Presidential Drive
Dayton, OH 45424
(937) 426-7800

Being just 2 miles away makes the Holiday Inn—Fairborn one of the most convenient hotels to the museum. It is, however, 24 miles from the Dayton International Airport.

Best Aviation Attraction

#89

U.S. Space & Rocket Center Museum

The **U.S. Space & Rocket Center Museum** in Huntsville, Alabama, has an interesting collection of rockets and space memorabilia. Much of it dates back to the Saturn era. My personal favorite is the instrument unit (IU). It is the brain of the Saturn V and controls its major systems. NASA at Marshall Space Flight Center (MSFC) in Huntsville home-managed the IU for the Saturn V. IBM's Federal Systems Division, where I once worked, contracted to build the IU. You'll see it and tons of other real space gear at this facility.

I am particularly impressed with the way the museum added a theme park feel to their facility. This integration of fun with learning makes a trip here a real treat. My favorite is the Space Shot. You'll be hurled 140 feet straight up in 2.5 seconds. Want to know what 4Gs feels like? This ride will teach you in a way you won't soon forget. This is very close to what the Apollo astronauts felt while riding atop the Saturn V. At the top of the Space Shot you'll experience weightlessness for 3 seconds. It's just long enough to give you an idea of what it must be like to be on the space station. What comes next is a thrilling 1G free-fall back to earth.

You may prefer Mars Mission, Apollo Cockpit Trainer, Lunar Lander, Kids Cosmos Energy Depletion Zone, or the Force Accelerator. All of these rides would be sellouts at Disneyworld. Everyone enjoys the realism seeing a movie in the Spacedome IMAX theater. You will too.

The **U.S. Space & Rocket Center Museum** is connected to the U.S. Space Camp, which is the number-one science camp for children in the United States. Spend a day at the museum and see what your child thinks of it. If you see a certain gleam in his or her eyes get out your credit card, the platinum one with the *high* charge limit, and sign 'em up. America needs a few more astronauts!

Dates

Open daily

Closed Thanksgiving Day, Christmas Eve, Christmas Day, New Year's Eve, New Year's Day

Hours

Daily 9:00 a.m.–5:00 p.m.

Admission Cost

Adults	$16.95
Children (3–12)	$11.95
Children (under 3)	Free

The price includes museum, rocket park, and one movie in the Spacedome IMAX theater.

Contact Information

U.S. Space & Rocket Center Museum
Huntsville, AL
(800) 63 SPACE
http://www.spacecamp.com
guestservices@spacecamp.com

G/A Airport Serving the Event

Huntsville, AL (Huntsville International—HSV)
Phone: (256) 772-9395
Tower: 127.6
ASOS: (256) 772-8074
Runways:

18R–36L	8000′ × 150′
18L–36R	10,000′ × 150′

FBO Name/Phone Number

Signature Flight Support
(256) 772-9341

Ground Transportation

Rental car

Best Place to Stay $$$

Radisson Suite Hotel Huntsville
6000 Memorial Parkway South
Huntsville, AL 35802
(256) 882-9400

#90

United States Army Aviation Museum

The **United States Army Aviation Museum** is located on Fort Rucker in south-central Alabama. It maintains a collection of over 160 military aircraft, including the largest collection of helicopters in the world. The exhibits trace the development and use of aviation by the U.S. Army in its key mission areas: troop and cargo transportation; observation, scouting, and liaison duties; medical evacuations; and the use of armed helicopters as an offensive weapon of the combined arms team. The public galleries represent the Army's involvement in military aviation from the beginning days with the Wright brothers up to the highly technological machines such as the AH-64 Apache and the UH-60 Blackhawk flown by Army aviators today. You'll see examples of each, including a full scale replica of the Wright Model B Flyer. The museum also tells the human side of Army aviation. The first fatality of an aviation accident was the Army's own Lt. Selfridge. Orville Wright was the pilot that day.

Army aviation began in 1909 with the Army's acquisition of its first Wright heavier-than-air flying machine. World War I saw the Army's aircraft strength grow from a few to more than 11,000. In August 1954 the Army Aviation School moved to Alabama. Thc first class began at what soon became Fort Rucker.

There is a lot to see here as you explore all that this museum and the military facility that houses it has to offer. Because it is on a functioning military base all visitors, ages 16 and older, must have photo identification. Vehicle operators must present a valid driver's license, vehicle registration, and proof of insurance in order to drive on Fort Rucker.

This is a difficult period for our nation and promises to be so for years to come. Depending on the Department of Defense security levels, access to Fort Rucker may be restricted at times. It is prudent to call the museum and ask about access restrictions to the installation prior to beginning your journey.

Dates

Open daily
Closed New Year's Eve, New Year's Day, Thanksgiving Day, Christmas Eve, and Christmas Day

Hours

Monday–Saturday	9:00 a.m. to 4:00 p.m.
Sunday	12:00 p.m. to 4:00 p.m.

Admission Cost
Free

Contact Information
United States Army Aviation Museum
P.O. Box 620610-0610
Fort Rucker, AL 36330
(334) 598-2508
curator@armyavnmuseum.org
http://www.armyavnmuseum.org/

G/A Airport Serving the Event
Enterprise, AL (Enterprise Municipal—EDN)
Phone: (334) 347-1211
CTAF-Unicom: 122.8
Runway: 5–23 5100′ × 100′

FBO Name/Phone Number
AT & S
(334) 393-2633

Ground Transportation
Rental car

Best Place to Stay $$
Holiday Inn Express Hotel & Suites
9 North Pointe Boulevard
Enterprise, AL 36330
(334) 347 2211

#91

Waddington International Air Show

Waddington International Air Show is the largest of the RAF's official air shows. The show is a favorite of visiting pilots and aircrew due to its excellent hospitality. Consequently, aircraft from around the western world turn up. Almost 100 aircraft were on static display at the airshow 2003. Seventy-one came from a dozen countries.

Airshow 2004 will be the 10th international airshow held at RAF Waddington since 1995. The Red Arrows will be flying their 40th season. The RAF **Waddington International Air Show** has been chosen as the official venue for the celebration.

RAF Waddington has become the center of operational excellence for air surveillance, strategic airborne reconnaissance, and related electronic warfare aspects, and this is to be the main theme for static aircraft visiting the air show in 2004.

Dates

June 26–27, 2004

Hours

Gates open	8:00 a.m.
Helicopter pleasure flying begins	9:00 a.m.
Air show begins	10:00 a.m.
Airshow ends	5:00 p.m.
Gates close	6:30 p.m.

Admission Cost

Adult	£15
Child (5–15)	£8
Child (under 5)	Free

Contact Information

Royal Air Force Waddington
Info@waddingtonairshow.co.uk
http://www.waddingtonairshow.co.uk/

Best Place to Stay $$$

Best Western Bentley Hotel & Leisure Club
Newark Road, South Hykeham
Lincoln, England
LN6 9NH
Great Britain
44 1522-878000

Best Place to Stay $$$

Best Western Grand Hotel
St Mary's Street
Lincoln, Lincolnshire, England
LN57EP
Great Britain
44 1522-524211

#92

Wheels & Wings Air & Car Show

Millville, New Jersey's **2004 Wheels & Wings Air & Car Show** is building on 4 years of successful air shows. Headlining the air show will be the United States Air Force Thunderbirds. This is their second trip to Millville.

Split Image Aerobatic Team, a hometown favorite, will lead the other acts. The USAF Reserve Command's Smoke-N-Thunder Jet Dragster will face off against the Hawaiian Jet Fire Truck. The Aeroshell Aerobatic Team promises to be a crowd pleaser this year as it has been in the past.

Night shows are catching on with the air show crowd nationwide and Millville will not be left behind. The Friday evening event will include a fireworks display.

Expect to see a variety of vintage and military aircraft along with aircraft from civilian aviation in the static display area. Ooops! I almost forgot to mention, this is also the largest *car* show in southern New Jersey.

Dates

May 1–2, 2004

Hours

Gate opens 8:00 a.m.
Event ends 5:00 p.m.

Admission Cost

Adults	$15.00
Children	$8.00

Contact Information

David Schultz Airshows
(856) 825-3272
http://www.schultzairshows.com/Millville2004.htm

G/A Airport Serving the Event

Millville, NJ (Millville Airport—MIV)
Phone: (856) 825-1244

Millville FSS: 123.65

Runway: 10–28 6002′ × 150′

Fly-in aircraft must be on the ground and parked no later than 0900L at the Millville Airport on Saturday and Sunday. All flight landings will take place on these days until 0900L by Millville FSS on 123.65 Mhz. Fly-in departures will be coordinated by the airboss on 132.95 Mhz for departure after the air show until 1700L each day.

Once you have landed at the airport, Millville FSS will hand you off to air show ground on 118.7 Mhz for parking instructions. Remain on this ground frequency for the remainder of your taxi unless instructed by the airboss.

FBO Name/Phone Number

Big Sky Aviation
(856) 825-3160

Ground Transportation

Rental car

Best Place to Stay $$$

Hotel Millville
545 N. High Street
Millville, NJ
(856) 825-9803

Best Place to Stay $$$

Best Western
1701 N. 2nd Street
Millville, NJ
(856) 327-3300

#93

White River Trout Fishing

Simply stated, **Gaston's White River Resort** in Arkansas is America's number-one trout fishing resort. The most important thing to know is that fishing the White River is allowed 12 months out of the year. November through March offers the best lunker trout catches; however, fishing is good all year. The water temperature is always at a trout pleasing 50 degrees. Everything you could need or want is provided to make your fishing experience top drawer. Many trophy-size brown and rainbow trout are caught and released here each year. One of them has your name on it!

The accommodations are varied. Everything from an easy-on-your-wallet motel room to a 10-bedroom villa is available. Each comes with one or more boats. Motors and everything else you'll need can be rented at the resort. A fly rod and reel is yours for $33 per day. Don't worry if you can't figure out which end is up, because knowledgeable guides are available! Their rates are reasonable. Count on spending $250 a day for two people with a shore lunch included.

Gaston's restaurant is worth the trip all by itself. Anyone wanting absolute top quality in every respect will discover that this is one of the nicest places they can go. The food is not inexpensive but is truly outstanding. You'll be seated in a big open dining room, built with large log cabin style timbers. It hangs out just a bit over the White River, so the views are stupendous. You can watch fishermen as they float down the fast-paced river, dragging their fishing poles. The popular tasty dish here is rainbow trout caught right out of the White River. The people, all of them, are extremely helpful and well trained. The entire operation is an example of what quality really means. People fly in from as far away as Houston, just to enjoy the Sunday brunch. Not to fish, not to sleep, not to enjoy the nature trails but just to eat. Gaston's is in the extreme northern part of the state—only a few miles from Missouri. The ride from Dallas in a Cessna 172 is a 2.5-hour flying adventure each way.

Gaston's is in the White River canyon. It is only 2 miles from the infamous Whitewater Land Development that haunted Bill and Hillary Clinton. While it isn't necessary to fly, it is a ton more fun and a whole lot quicker.

Dates

Open year round

Website

www.gastons.com

Contact Name

Danny Gaston

Telephone Number

(870) 431-5203

E-mail Address

gastons@gastons.com

G/A Airport Serving the Event

Lakeview, AR (Gaston's Airport—3M0)

Alt: 479ft

Lat: 36.20.92N

Lon: 92.33.43W

Phone: (501) 431-5202

CTAF-U: 122.8

Runway: 3200′ × 75′ 06–24–Turf

Fuel: Air BP

It is easy to find even without a GPS. Simply fly to the Flippin VOR inbound to the west end of Baxter County Airport and look off your left wingtip down into the White River Valley. Gaston's Airport, the prettiest 3000-foot-plus grass strip you can imagine, will come into view. Two relatively high ridges curve around both ends of the runway. If you are not a little on the adventurous side, landing at Baxter County Airport in Mountain Home is a great alternative. It's an exceptionally nice airport and Gaston's will gladly come pick you up. It's only about a 15-minute drive.

Gaston's Airport was built back in the days when most people that flew in came in Tri-Pacers, 182s, Bonanzas, and small conventional gear aircraft. Today corporate-owned King Airs arrive regularly.

The airport is exceptionally well maintained but you need to really pay attention on the approach because the trees that line both sides of the runway are now very mature and are reasonably close to the runway on both sides. A carrier-type approach is required to some degree, making it more difficult than your average country airport. Because of the trees, you won't be able to see the entire runway, until you're on final and "in the slot." Most people land to the west. A light tailwind will add to your ground speed producing some surprising results. If the grass is a little wet braking action is reduced.

Keep a sharp eye out for Valley Airpark and Flippin traffic while on downwind for runway 24 at Gaston's. When transitioning to base leg, scan to the right to check for traffic inbound to Baxter County. It is served by airlines so there are likely to be some fast movers out there. Also scan for aircraft inbound to Gaston's on a straight-in approach. Finally, when turning final, keep a sharp eye out for departing traffic. Once you get below tree level your choice of evasive action is limited to down or up.

Best Aviation Attraction

#94
Wings Over Houston Airshow Festival

Air show excitement was invented at Houston's Ellington Field when **Wings Over Houston** performers introduced the **Tora, Tora, Tora** show several years ago. It continues to delight local crowds and has become a featured act at select air shows across America. In 2004 the Air Force will send the Thunderbirds to heat up the skies over Houston with a flight demonstration that only they can accomplish. It is no wonder that this annual event has become one of the top five air shows in the country and the best in the southwest!

If you are a pilot, don't miss it. This is the only large air show where you are invited to land your aircraft and enjoy the show. Plan to arrive well before it begins and depart well after it concludes.

Houston is as much about performance as it is aircraft. *Tora, Tora, Tora* is a reenactment of the sneak attack on Pearl Harbor. The Zekes and Zeros swing in out of the rising sun with their guns blazing. You'll hear them loud and clear and you'll be very grateful that they're firing blanks. Without warning they'll simulate dropping a load of 500 pounders just across the runway from where you'll be standing. The roar is deafening. This is the same sound that awakened our sailors on December 7, 1941. You'll see the explosions and stand frozen as the burning fuel oil billows into the sky in front of you. Thank God this is only an act and may it never really happen again!

No contemporary air show would be worth attending without the support of the United States Armed Services. Houston is no exception. The top military hardware are shown in flybys, static displays, and aerial performances. These are perilous times for our country and all of our military might could be called for duty. As it stands now you'll see a B-1B and a B-2 flyby. Both will rock you back, especially the B-2. It is really special.

The Navy will have an F-18E Super Hornet in a static display for you to see. It is huge! Its F-14 Tomcat will put on a show. This is the Navy's single pilot fleet protector and the pride of the Blue Angels.

The Air Force will answer with its F-15 Eagle, the current air superiority fighter. Yes, the Raptor will soon deploy to take over that role. For today, the F-15 is the top dog. This aircraft can climb straight up at a vertical speed exceeding 30,000 feet per minute!

Dates

October 16 and 17, 2004

Hours

Saturday	8:00 a.m. to 6:00 p.m.
Sunday	8:00 a.m. to 5:00 p.m.
Air show	10:00 a.m. both days

Admission Cost

Adult	$15.00
Child	$5.00

Contact Information

Wings Over Houston Airshow Festival
(713) 266-4492
info@wingsoverhouston.com
www.wingsoverhouston.com

G/A Airport Serving the Event

Houston, Texas (Ellington Field—EFD)
Phone: (713) 847-4200
Tower: 126.05
ATIS:
Runways:

17L–35R	4600′ × 75′
17R–35R	9000′ × 150′
4–22	8000′ × 150′

Air show attendees flying their own planes to Ellington Field for the air show must be on the ground and in place by 9:00 a.m. on show days. If aircraft are not at least in the pattern on these days, they may be waved off to another airport. During the show, the field will be closed to all air traffic.

Pilots will be instructed where to park upon arrival. Everyone arriving in the aircraft must have an air show ticket or be prepared to buy one.

Aircraft will not be able to leave until Ellington Field is reopened to air traffic after the show concludes.

FBO Name/Phone Number

Southwest Airport Services

(281) 484-6551

Ground Transportation

Rental car

Best Place to Stay $$$

Hilton Houston NASA Clear Lake

3000 NASA Road One
Houston, TX 77058-4322
(281) 333-9300
(281) 333-3750 (fax)

The Hilton NASA Clear Lake is a newly renovated, suburban resort overlooking Clear Lake. It is minutes from the gates of NASA's Johnson Manned Space Flight Center and only a short 5 miles to Ellington Field. Don't come to the air show without planning a visit to NASA's visitor center.

Best Place to Stay $

Comfort Inn NASA/Houston

750 W. NASA Road One
Webster, TX 77598
(281) 332-1001
(281) 554-2967 FAX

Best Aviation Attraction

#95

Wright Brothers National Memorial

The first successful sustained powered flight in a heavier-than-air machine was made here by Wilbur and Orville Wright. That is such a simple sentence to craft and such a difficult feat to accomplish. From the beginning of human history, virtually every person has looked skyward and wanted to take to the air. Each had wondered what it would be like to fly. In 1932, a 60-foot granite monument was placed atop the 90-foot tall Big Kill Devil Hill to commemorate what had happened here on December 17, 1903, when *man's dream had been realized!*

Start your visit at the visitor center. It is packed with exhibits that you'll want to see. Don't miss the program that is given by a park ranger in the auditorium. It is stirring.

Next tour the reconstructed camp buildings; they will give you a glimpse into how harsh conditions must have been for the Wrights. There are two buildings. One served as a hangar for their aircraft, The Wright Flyer. The other was a workshop and living quarters. Each is filled with items similar to the ones used by the Wrights.

You will notice a large granite boulder. Walk over to it and read the inscription. This is the spot where the Wright Flyer with Orville at the controls lifted off. This is the spot where man's imagination of what it might be like to fly became reality. In a straight line extending from this spot you will notice four number markers. They indicate the distance of each of the four flights made that day. The longest is 852 feet from where you'll be standing. The final and longest flight of the day lasted 59 seconds and reached an altitude of 10 feet.

The First Flight Airstrip sits in the trees to the west of where you'll be standing. It is only a short walk. If the National Park Service has found a new contractor by the day of your visit, you'll be able to take an aerial tour of Kitty Hawk in a small airplane. You may, of course, land here in your own aircraft. If you're a pilot you'll certainly want to do that. It is a short runway at an airport which offers no services at all.

The museum and the grounds can comfortably be toured in 2 hours.

Dates

Open every day except Christmas Day

Hours

9:00 a.m.–6:00 p.m. during the summer
9:00 a.m.–5:00 p.m. the rest of the year

Admission Cost

$3 for 7 days
$10 Annual

Contact Information

Wright Brothers National Memorial
(252) 441-7430
http://www.nps.gov/wrbr/index.htm
caha_interpretation@nps.gov

G/A Airports Serving the Event

Kill Devil Hills, NC (First Flight—FFA)
Phone: 252-473-2600
CTAF-Unicom: 122.9
Runway: 2–20 3000′ × 60′

There is no FBO at this airport.

It is only a short walk to the Wright Brothers Museum. A word of caution—you are only allowed to tie down for a 24-hour period. You may not tie down for more than 48 hours total in any 30-day period. No ground transportation is available from this airport.

Manteo, NC (Dare County Regional—MQI)
Phone: (252) 473-2600
CTAF-Unicom: 122.8
AWOS: 128.275
Runways:
5–23 3000′ × 75′
17–35 4300′ × 100′

FBO Name/Phone Number

Dare County Airport Authority
(919) 473-2600
info@flymqi.com
www.flymqi.com

Dare County is a full-service airport with several instrument approaches. Fuel and repairs can certainly be found here. Be warned though, radar coverage is not reliable below 4000 feet.

Ground Transportation

Enterprise Rent A Car
800 Rent A Car

Best Place to Stay $$$$$

Kitty Hawk is on North Carolina's outer banks. There are miles and miles of ocean front beach houses available for rental. It is rare to find a beach house available for less than a week.

The better accommodations command rentals of $8000 a week during the sought-after summer period. This is a lovely spot to spend time if you want to be on the beach and away from it all. Here you will be. It is very remote.

Finding just the right one is difficult. It is best to first find a local realtor who deals with rentals. A good place to start is:

Prudential Resort
Reservations 1 (800) 458-3830
info@resortrealty.com
http://www.resortrealty.com/book/query.html

They have local offices in each of the outer bank towns, Corolla, Duck, Kitty Hawk, and Nags Head.

Best Place to Stay $

Holiday Inn Express Kitty Hawk
3919 N. Croatan Hwy.
Kitty Hawk NC 27949
(252) 261-4888

The Holiday Inn Express has an outdoor swimming pool and is a short walk to a life guarded beach. All 98 rooms are spacious and have cable TV, telephones, and refrigerators. All rooms were renovated in 1997. The hotel provides a complimentary continental breakfast for guests each morning in the lobby. Nonsmoking and handicapped-accessible rooms are available. Year-round group rates are available.

Best Aviation Attraction

#96

Yankee Air Museum

The **Yankee Air Museum** is located in an historic hangar at Willow Run Airport in Ypsilanti, Michigan. Its entrance is guarded by the imposing shape of a giant B-52. Housed in a 1941-era hangar which played a role in the production of nearly 8700 B-24 Liberator bombers, it brings honor to the aircraft of war and the brave men they carried into battle. The Yankee Air Force which supports this museum has members throughout the United States and the world. The membership is well over 3000 and growing.

The **Yankee Air Museum** consists of an indoor aircraft display, an outdoor aircraft display, and numerous rooms devoted to specific aviation themes or time periods. Attention is drawn to aviation in the World War I, World War II, Korea, and Vietnam periods. Women in aviation and the B-24 Liberator bomber are other subjects of this studious collection.

Many of the historic aircraft housed here are flyable classics. Others are on static display. You may view skilled mechanics and restoration crews actively involved in the process of slowly bringing other aviation treasures back to life.

The flyable aircraft of the museum are frequently invited to participate in air shows across the country. The best show of the season for the YAF is always the YAF Memorial Day Open House. This is the best time to visit.

Be certain to schedule a FLEX (Flight Experience ride on their B17, *Yankee Lady*) whose cost is $400 for a flight of approximately a half hour. The flight crew of three is joined by up to seven first-time aircrew members on each excursion. You really do want to do this!

Aviation is about flying, not sitting on the ground, and the B-17 is the epitome of air power in World War II. Without it we would probably all be speaking Italian, German, or Japanese. To really understand what it was like to be a member of a World War II combat aircrew you need to take a ride in the machine that they risked their lives in. As you take off, let your imagination return you to those days when the hopes of democracy rode on these wings. Think about the mission ahead. Look out either waist gunner port, they're huge aren't they? Now imagine what it must have been like to see a ME-109 screaming down from the clouds above with puffs of smoke coming from each wing. Imagine how cold it must have been at 25,000 feet on a winter's day above Berlin in this unpressurized, unheated battlewagon. Do this one time and you will understand why this national organization and its museum work so hard to preserve these aircraft and these memories. **"Those who forget their history are destined to repeat it."**

Dates

Open daily
Closed Mondays and holidays

Hours

Daily	10:00 a.m.–4:00 p.m.
Sunday	12:00 p.m.–4:00 p.m.

Admission Cost

Adults	$7.00
Seniors	$5.00
Children	$3.00

Contact Information

Yankee Air Museum
Willow Run Airport
Ypsilanti, MI
(734) 483-4030
yankeeairmuseum@provide.net
www.yankeeairmuseum.org

G/A Airport Serving the Event

Detroit, MI (Willow Run Airport—YIP)
Tower: 125.275
ASOS: 734-485-9056
Runways:

5R–23L	7526′ × 150′
5L–23R	6655′ × 160′
9R–27L	6511′ × 160′
9L–27R	7294′ × 160′
14–32	6914′ × 160′

FBO Name/Phone Number

Willow Run Jet Center
(734) 483-3531
http://www.jetctr.com/

The Yankee Air Museum is adjacent to ample tie-down space and Jet Center East making it a great fly-in destination.

Ground Transportation

Rental car

Best Place to Stay $$$

Ypsilanti Marriott at Eagle Crest
1275 S. Huron Street
Ypsilanti, MI 48197
(734) 487-2000

The Ypsilanti Marriott at Eagle Crest is the only hotel in southeastern Michigan with an 18-hole championship golf course and GPS system. Live life to the fullest, plan your Memorial Day outing here. Enjoy the resort setting, play golf, take in the YAF Memorial Day Open House, tour the museum, take a ride on the *Yankee Lady*! Live large! Smile big!

Best Aviation Attraction

#97

You the Pilot

You bought this book because something inside you stirs a little bit when a plane passes overhead. Unlike others, you look up. Over the years you've studied airplanes just enough to be able to identify them as they fly above you.

Your love of planes stretches way back to your childhood. Your family trips to the airport were irregular. You didn't go to take a trip. The airport was the trip. Your dad would park just at the end of the runway on the street side of the fence. A few other cars would be there also but not many. Then it would happen. You would hear it coming from a distance and then you would see it. As the approaching plane drew nearer the sound reached such a crescendo that you had to cover your ears as did your brother. Do you remember?

Planes have always been special to you. While others go white-knuckled at the thought of transcontinental business flights, you sit with your nose glued to the window watching the earth pass by beneath you and hoping the flight will last just a little longer. You put on the headsets hoping there will be a channel which lets you listen to your flight deck crew talking with air traffic control.

You've read all the books, seen all the movies, been to all the museums and air shows. You have been to the Cessna Pilot Center and taken the Discovery Flight. Life is moving along. Now it is time to fish or cut bait. Will you become a pilot or forever be a dreamer? When the winter days of reflection come will you say "I wish, I shoulda' coulda'" or "I'm glad I did"? Will you sing the song of regret or watch the movie of memory? What do you say? Is today the day, and if not, how many more tomorrows must come?

If this is your day, let me help you along. It is going to cost about $6000 to get your license, a little more perhaps, a little less maybe, but about $6000. That's not so bad really. The best part is that you pay as you go. The $6000 will dribble from your wallet a little at a time. The next best thing is this: once you solo and you're on your own the leash gets a little longer the further you advance. When your solo cross-country is complete, your instructor can free you up to make some pretty interesting cross-country solo flights. All you have to do is budget enough money for the next flight.

You don't even need to decide to finish, just to get started. You'll spend about 12 hours with your instructor before he turns the plane over to you for your first unassisted trip around the patch. Then you'll know.

Get started today with **Microsoft Flight Simulator 2004.** Take the Rod Machado instruction. Let me assure you it is a lot harder to fly the simulator than it is to fly the airplane. If you can satisfy the automated Rod you can easily satisfy any skin-and-blood flight instructor.

Where will you take your lessons? The best answer is that it depends. A Cessna Flight Center is a good choice but it can often be more expensive than nearby competitors. Think about the kind of flying you'll be doing once you have the license and be guided by that. If you intend to take a lot of cross-country trips, it is best to train at a busy controlled field. This will help you get used to talking with controllers, listening to all the radio chatter from other pilots, and fit in with the traffic flow. It will make you comfortable. If you train at an uncontrolled country field, it will be difficult for you to gain comfort when forced to land at big city controlled fields. But if you intend only to go up on pretty days and play Don Quixote with the puffy white clouds why bother yourself with the high prices and anxiety of a big city airport.

Before you solo you'll need to get a third-class medical ticket from an FAA certified medical examiner. Before you take your checkride you'll need to pass the written.

My advice on the written is to buy a computer-aided study course from John and Martha King. Take a look at their products on line at http://www.kingschools.com/. You can find others that are less costly but you won't find any that are as valuable. Their training methods work and they cover *all* the material.

I am excited that you have decided to become a pilot and I hope that you'll write me at jpurner@flyingsbest.com and let me know how you're coming along. You see, I really love flying and I like knowing that other people share my enthusiasm.

Good luck and good flying!

Best Aviation Attraction

#98

Young Eagles

(*Courtesy EAA.*)

Over a million girls and boys have become **Young Eagles.** The one millionth **Young Eagle** attained that special status on November 13, 2003. It took an army of more than 85,000 volunteers who have supported the program for a decade to make this happen. Now it is time for that special youngster you know to do the same.

The mission of the **Young Eagles** program is to introduce young people, ages 8 to 17, to the world of aviation and provide meaningful flight experience—free of charge. It took wings on July 31, 1992. Under the skillful leadership of Steve Buss, executive director of the **EAA Young Eagles,** the program continues. **Young Eagles** has become a part of the very fabric of EAA. The mission is too significant to stop now. It is important to involve young people early to assure aviation's future. A significant number of former **Young Eagles** have earned their pilot's license and have returned to fly their own **Young Eagles,** completing the journey started with their initial flight. The EAA is committed to providing the tools and support our **Young Eagles** desire, as they explore all the world of aviation is able to offer them. Today's **Young Eagles** will build aviation's next century.

This attraction takes place at hundreds of America's small and large general aviation airports each weekend.

"Tower—Young Eagle Flight ready for takeoff, runway 36." When an EAA member pilot Flight Leader speaks these words, another young American is seconds away from feeling the joy of flight for the first time. It's safe. It's fun. It's free. And it's fantastic! Becoming an EAA Young Eagle is as easy as one-two-three!

Step One

Contact the **Young Eagles** office at 1 (877) 806-8902.

Step Two

They'll tell you where to find a pilot near you!

Step Three

Take off!!! You and your personal pilot will zoom through the skies!

Anyone between the ages of 8 and 17 can be an EAA Young Eagle just like that! At the end of this no-cost flying adventure, an official **Young Eagle** certificate will be issued and the name of this **Young Eagle** will be entered in the world's largest log book.

Dates

Continually but primarily on weekends

Hours

Daylight

Admission Cost

$0—Zero—Nada—Nothing—Free

Contact Information

Young Eagles
P.O. Box 2683
Oshkosh, WI 54903-2683
(920) 426-4831
(877) 806-8902
(920) 426-6560 (fax)
yeagles@eaa.org
http://www.youngeagles.org.

G/A Airport Serving the Event

All of them!

#99

Zeppelin Museum—Friedrichshafen

The **Zeppelin Museum—Friedrichshafen** is the world's greatest repository of the history of airship flight and technology. Its principal focus is on the work of Ferdinand Graf von Zeppelin and the aircraft that bear his name. He began the development of rigid airships in 1890. His work is being carried on today by ZLT Zeppelin Luftschifftechnik GmbH & Co. KG.

The museum is housed in the former Harbor Railway Station located on the promenade of Lake Constance. There are two areas of interest for this museum. The first is the technology of airships. These exhibits range from the very beginning of airships up to the construction of the current model, the Zeppelin NT. The second exhibit interest is art. The extensive collection shown on the upper floor demonstrates works of art of the Lake Constance region from the Middle Ages to the present day.

The airship collection of the Zeppelin Museum is divided into six major departments. Together they bring all aspects of the history and technology of airship flight to life. The museum's main attraction is the reconstruction of a 100-foot-long part of the LZ 129 *Hindenburg*. Walking through this exhibit gives visitors the experience of the great dimensions of the "Giants of the air." The six technology departments are:

Airship construction types, physical principles, takeoff, and landing
Airship construction, motors, cars, and aerodynamics
Navigation and radio technology
The history of the civil use of airships including the Zeppelin NT
The history of the military utilization of airships
The history of the Zeppelin company, Luftschiffbau Zeppelin GmbH, and its succeeding enterprises

The art collection contains an extensive collection of works by Otto Dix, Max Ackermann, Karl Caspar, and Maria Caspar-Filser. The art department presents paintings and sculptures by local artists from the Lake Constance region ranging from the Middle Ages up to the present day and paintings by Karl Caspar and Maria Caspar Filser.

After a day of exploring the history of airships and marveling at the unique possibilities this type of aircraft provides, you will long for a ride in one. This wish is easily accommodated. You can go aloft in a state-of-the-art Zeppelin NT.

Book this adventure well in advance. A modern airship can only accommodate 10 passengers. Make your reservation by contacting:

Zeppelin Luftschifftechnik GmbH
Press and Public Relations:
Dr. Jeannine Meighörner
Zeppelin Werftgelände 31
88045 Friedrichshafen
Germany
Tel: 0049-7541-202-577
Fax: 00049-7541-202-499
e-mail: jmeighoerner@zeppelin-nt.com
http://www.zeppelin-nt.com/index.html

Museum Dates

Open Tuesday through Sunday
Closed Mondays except public holidays

Museum Hours

May to October	10:00 p.m. to 6:00 p.m. (admission until 5.30 p.m.)
November to April	10:00 a.m. to 5:00 p.m. (admission until 4:30 p.m.)
During AERO (The International Flight Trade Fair)	10:00 a.m. to 8:00 p.m.
Easter Monday	10:00 a.m. to 5:00 p.m.
Whit Monday	10:00 a.m. to 6:00 p.m.

Admission Cost

Adults	€6,50
Seniors	€5,50
Children (ages 6–16)	€3,00
Family ticket	€13,00

Contact Information

Zeppelin Museum
Seestraße 22
88045 Friedrichshafen
Germany
49 / 7541 / 3801-0
49 / 7541 / 3801-81 (fax)
http://www.zeppelin-museum.de/firstpage.en.htm

G/A Airport Serving the Event

Flughafen Friedrichshafen GmbH
49 (0) 7541 / 284-01
49 (0) 7541 / 284-119 FAX
info@fly-away.de
http://ww2.fly-away.de/en/

Ground Transportation

AVIS
Reinbold GmbH
Agentur der Avis (Autovermietung GmbH & Co. KG)
49 (0) 7541-930700
49 (0) 7541-930705 (fax)
http://www.avis.de

Getting There from the United States

Without a doubt, the most convenient and interesting access to this museum is through the Zurich International Airport. While there are airlines offering connecting flights to Friedrichshafen, don't take one. Instead go directly to the train station which lies just under the airport terminal. Take the train to Romanshorn. From Romanshorn you travel across beautiful Lake Konstance on a ferry to Friedrichshafen. The ferry terminal is a short walk from the museum.

Best Place to Stay $$$

Best Western Hotel Goldenes Rad

Karlstrasse 43

Friedrichshafen, 88045,

Germany

0049-(0)7541-28 50

0049-(0)7541-285 285 (fax)

This hotel is located exactly in the town center. It is a short walk to the lakefront and the museum.

Best Aviation Attraction

#100

$100 Hamburger

Pilot lore has it that a hamburger costs $100. That breaks down to 5 bucks for the burger and $95 for the flight to and from the airport restaurant where it is sold. Pursuit of the $100 hamburger is glue that holds most of general aviation together. You've seen it. A group of airport rats are hanging around the local 'port on any given Saturday. After hangar-flying the morning away they begin to get a "rumbly in their tumbly" as Pooh Bear might say. These guys are hungry. Somebody says I know this great place we can fly to for lunch. A debate ensues about just where the best-yet airport-obtainable burger is. The issue is decided and 14 airplanes fill the sky as this group makes a beeline for a faraway lunch.

To the uninitiated this seems frivolous; to any aviator it is essential. To be able to use an aircraft when you really need to, on a business trip, your skills need to be up to snuff. The only way to do that is to fly. Circling the patch gets plenty old, plenty quick and isn't that helpful anyway. To be mission ready, you need to do cross-country flying to places you have seldom or never been to before. Burger runs are really training flights with a reward in the middle.

Strange as it may seem I always file IFR on my burger runs, even on CAVU days. I like to get into the "system" for two reasons. First I need as much practice as I can get, particularly flying the ILS approach into my home field. Next, it is safer. The controllers do a first rate job of separation. If an airplane that you do not see comes into your flight path, the controller is going to give you a heads-up. I also practice using every piece of equipment in my panel. I don't want to wait for a dark stormy night to try to figure out the GPS.

Another good purpose of the **$100 Hamburger** flight is to promote general aviation. Take a neighbor along on a Saturday or a client on a Thursday. Best of all take your wife to a romantic dinner in a nearby garden spot on any Friday night and be a hero for at least a week. Don't tell her where you're going; just make it special.

A pilot can find an up-to-date list of the best 1675 fly-in restaurants in the United States online at www.100dollarhamburger.com or by buying the book *The $100 Hamburger* at most pilot shops or at http://www.avshop.com. Yes, it is also available at amazon.com and many commercial booksellers.

If you and your clients are more into golf than burgers, check out www.flyingolf.com or buy the book *The $500 Round of Golf* at the same places that sell the big-buck burger book. It will guide you to hundreds of fly-in golf courses scattered around the country.

It would be shameless of me not to mention that I wrote those books and manage those Websites. I did it initially because I wanted to learn where more fly-in restaurants were located and wanted to share my findings with other pilots and report their feedback to our growing community. The burger-run for me was initially only about fun. It was later that I realized it was more about training. Finally, I became aware that it was also about promoting general aviation.

The story of my book *The $100 Hamburger* and the pilots that follow it has been reported in newspapers throughout the world, including large newspapers like the *Chicago Tribune,* the *Boston Globe,* and the *Wall Street Journal,* and small newspapers like the one in your neighborhood. It has been the focus of television and radio and has heightened the awareness of the nonpilot community about the utility of general aviation aircraft. We all need to do everything we can to portray general aviation in a positive light in this period after 9/11. Light airplanes are subject to countless flight restrictions for mostly irrational reasons. It was not light aircraft but heavy airliners that brought down the World Trade Center towers. Those of us that fly know very well that you can haul a lot more explosives in the family car than you can get off the ground in a Piper Cherokee. General aviation airports are being closed (Meigs Field) and TFRs are being placed over private places (Disneyworld) by political opportunists hiding behind a shield of national security. It is up to us to tell the other side of the story. It is up to us to show the utility of the planes we fly and remove their danger by making them familiar.

Best Aviation Attraction

#101

1940 Air Terminal Museum

When I travel by commercial airliner these days, I often feel as though I have been herded onto a cattle car or a Greyhound bus. The gloom sets in as I arrive at the airport parking garage. The architecture of the terminal is efficient and foreboding. Huge and sterile, it's designed to move passengers like packages through a FedEx processing. Commercial air travel is today an unpleasant experience which becomes worse the further into the belly of the beast you go. It isn't even worth mentioning the knee-knocking confinement of the seating arrangements or the cardboard that passes for food.

Air travel wasn't always this way. There was a time when it was an adventure to be looked forward to with zeal. The food service rivaled the best restaurants. As a matter of fact, there was a time when airlines competed based on the quality of their food service not their fare structure. Continental was the home of the "proud bird with the golden tail" and they offered "champagne flight." The experience began as you drove up to the airport terminal. It was beautiful, inside and out.

The **1940 Air Terminal Museum** is an attempt to reconnect us with the golden age of commercial air travel. It is a beautiful and rare example of classic art deco airport architecture from the golden age of flight. This terminal actually served Houston during the years when air travelers dressed in their finest and embarked for exotic destinations aboard roaring propliners like the Douglas DC-3 and the Lockheed Constellation. This terminal had been sitting vacant and in disrepair for many years when the farsighted Houston Aeronautical Heritage organization bought it. They have spent years and a ton of cash bringing it back to its former glory. The museum is a work in progress as only the north wing is opened. Many period displays and exhibits are provided to bring the moment to life. Aircraft suitable to the museum's history are available for inspection on the ramp.

The best way to see this museum is through the prism of experience. Book a Southwest flight from anywhere to Houston's Hobby Airport. It is one of the first two airports served by this contemporary model of efficiency. The other is Dallas' Love Field. Once you are off-loaded, grab a cab for the short ride to the other side of the field or call the museum for transportation guidance. They may have a shuttle from the active side of the field by the day you make your visit. The contrast in service and style will strike you instantly. The hardest part of the trip will be getting back aboard the return Southwest flight.

Dates

Open daily
Closed major holidays

Hours

10:00 a.m. to 4:00 p.m.

Admission Cost

Adults	$2.00
Children	$1.00

Contact Information

1940 Air Terminal Museum William P. Hobby Airport
8325 Travelair Road
Houston, TX 77061
(713) 454-1940
http://www.1940airterminal.org/
zalar@1940airterminal.org

G/A Airport Serving the Event

Houston, TX (William P Hobby—HOU)
Phone: (713) 640-3000
Tower: 118.7
AWOS: (713) 641-21 22
Runways:

17–35	6000′ × 150′
4–22	7602′ × 150′
12R–30L	7602′ × 150′
12L–30R	5148′ × 100′

FBO Name/Phone Number

Million Air-Houston
(713) 641-6666

Ground Transportation

Rental car

Best Place to Stay $$$$$

The Magnolia Hotel
1100 Texas Avenue
Houston, TX 77002
(713) 221-0011
http://www.themagnoliahotel.com

This is a historic downtown Houston hotel which has recently been restored to its former glory. It is the fitting retreat for your visit to the **1940 Air Terminal Museum.**